# grasslands

MICHAEL P. BERMAN

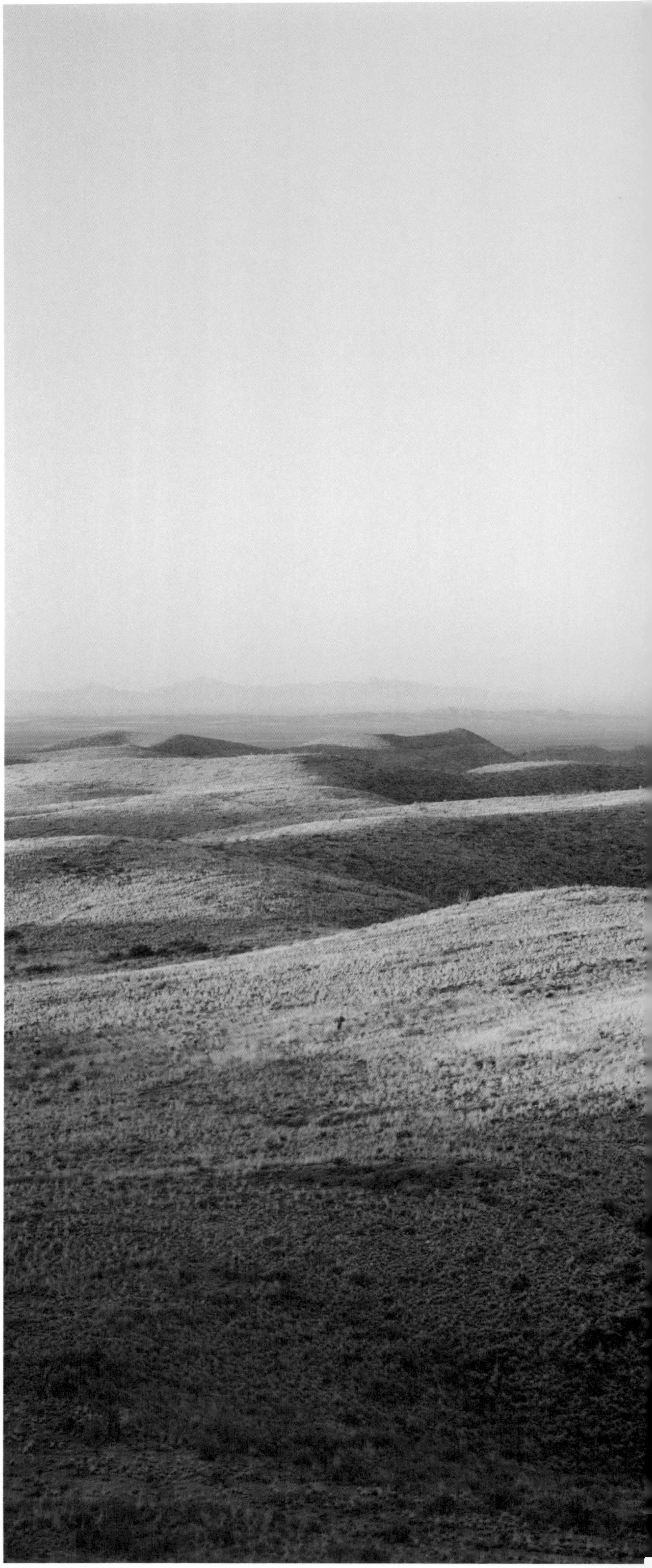

**Rough Canyon,** Otero Mesa, New Mexico

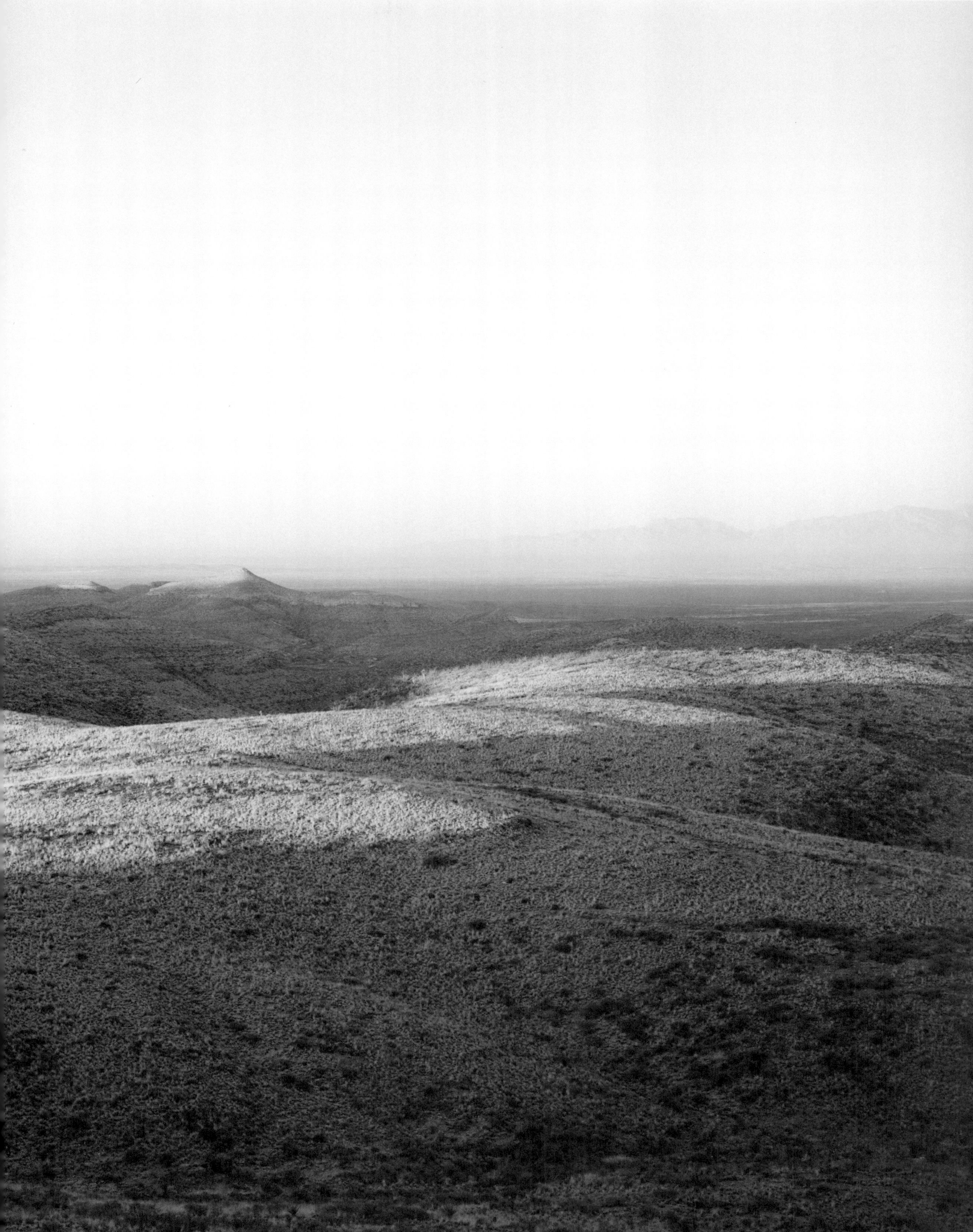

**MICHAEL BERMAN'S CHIHUAHUAN GRASSLANDS**

BY WILLIAM DEBUYS

**EVERY PLACE NEEDS A POET**, someone to reveal the heart of the land. The poet of place tells us where we are, and he also tells us why the particular whereness of a place is important. He shows us the intricate dance of space and time written on the land, which is everywhere different, and he shows us how that dance shapes our relationship, not just to the world that is, but to the world that should be. He helps us hone our senses and our sensibilities so that we use both our hearts and our minds to apprehend our surroundings in sharper focus. Sometimes the poet of place does all this wordlessly; sometimes his poems are pictures.

Michael Berman is such a poet. The place he would have us know is called the Chihuahuan Desert, but immediately we encounter an obstacle and it is not of his making. The word *desert* draws its power from the extremes it describes: Saharan wastes, the austerity of the Holy Land, the baroque flora of the Mohave. *Desert* fails to prepare us for the arid prairies of the northern Chihuahuan biome—and we generally ignore what we lack the wherewithal to understand. We hardly see what is there, and still less do we see what used to be there. As Berman puts it, "The Chihuahuan Desert is of little value to the modern culture without an aesthetic for grasslands."

The aesthetic and cultural roots of most Americans lie in forested lands. Grasslands by contrast have tended to be a puzzlement. In the settling of the continent by neo-Europeans, grasslands were generally something to be got past on the way to somewhere else, somewhere with trees. The grasslands of the Chihuahuan Desert attracted few westering Americans until most other places were filled in. They were also among the last lands to be settled in Mexico. People found them difficult and intimidating, and they still are. They also found the native people of the grasslands—the Apache—extremely difficult and intimidating, and in a way, although the Apache are largely gone, the people who live in the region today still hold their own in that regard, too, at least the ones who carry guns, which includes a lot of them.

Maybe it was the quality of intimidation, not just the beauty of the landscapes, that drew Berman to the Chihuahuan grasslands. They demand strength and persistence and loyalty. They turn away the insincere, and they require much even from their admirers, perhaps especially from their admirers. They range from tawny seas of grass where antelope skitter and bound to ghost grasslands where, for a host of reasons linked to the behavior of cow-tending humans, various shrubs like mesquite and creosote have supplanted the once-dominant grasses. You see a lot of the past in the present of the grasslands. You see the bones of an older world.

The Chihuahuan desert grasslands are a region, where, in Berman's words, "things fall apart." Geographically in northern Chihuahua the skeletal logic of great mountain chains—both the Rockies and the Sierra Madre Occidental—gives way to an archipelago of choppy volcanic blocks and sawtooth pinnacles. The scattered mountains seem to float on a sea of sweeping plains. It is a land almost impossible to see

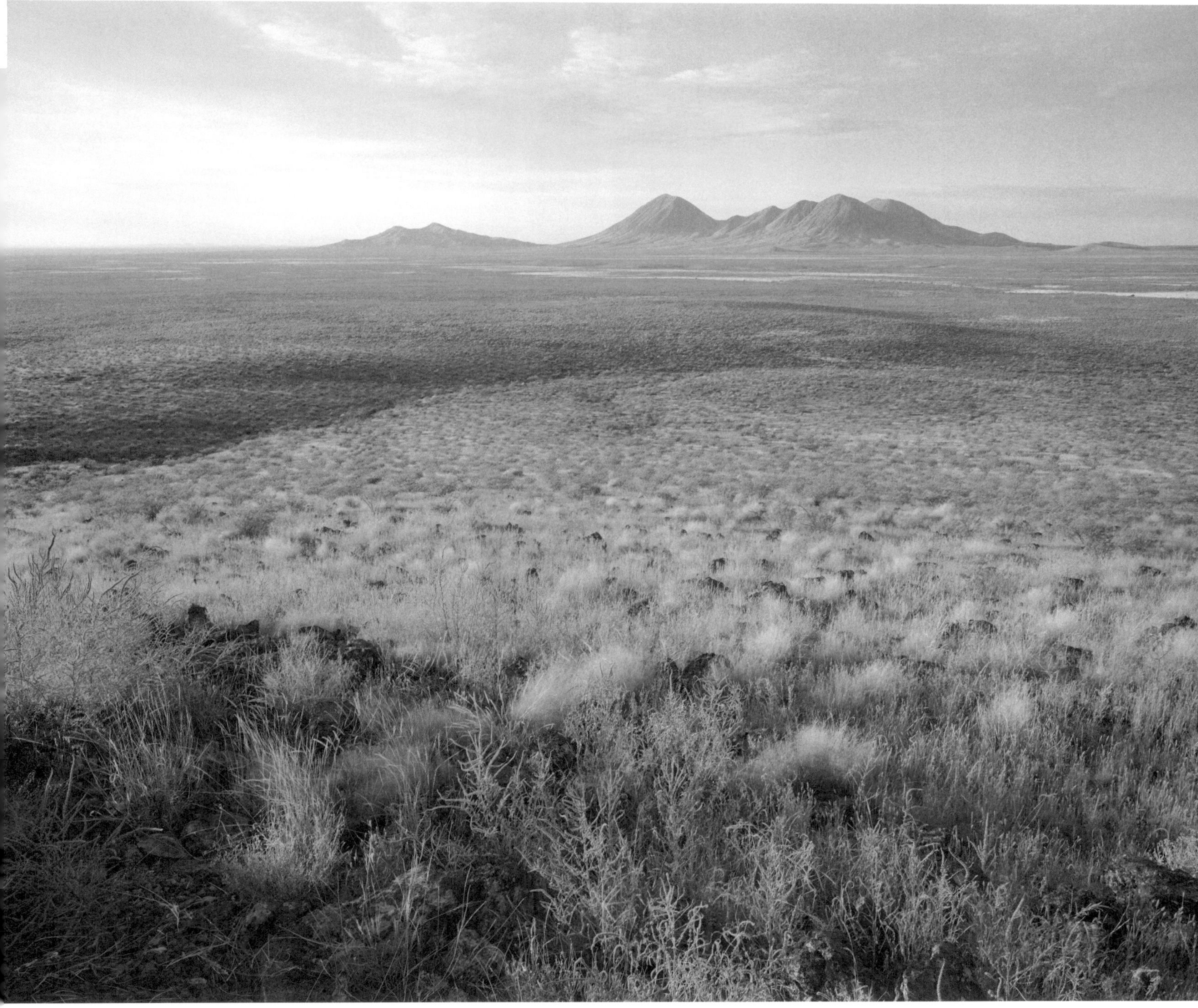

**Mt. Riley**, New Mexico

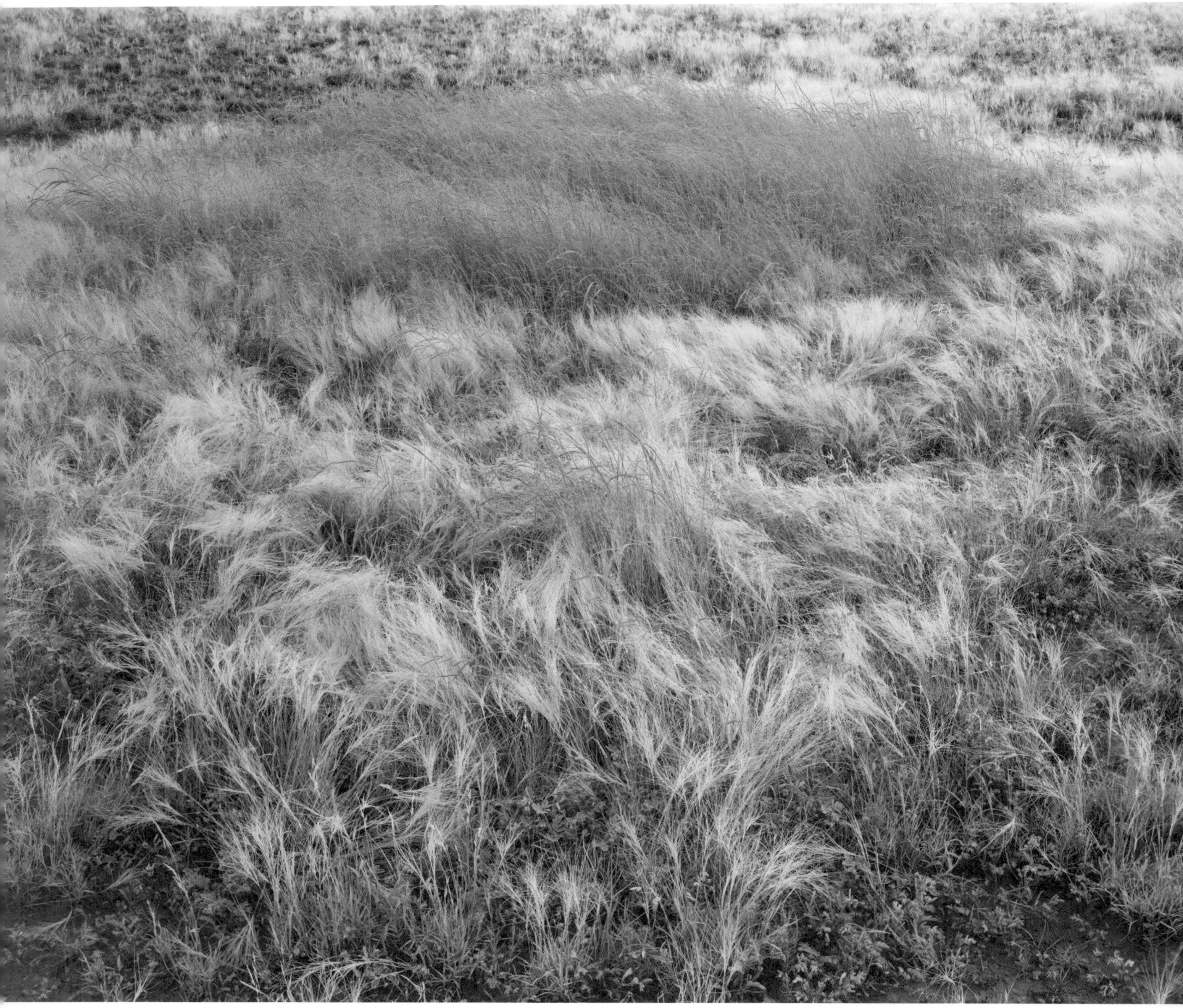

**Burrograss**, Jornada del Muerto, New Mexico

in the full light of day when the sun dissolves distances and melts the edges of things. It is a land that most reveals itself at dawn and dusk when shadows are long and the creatures of the desert slink about.

Part of what falls apart in the Chihuahuan Desert Grasslands is the civil relationship between two nations. The region is bisected by the international border between the United States and Mexico, easily the most pathological portion of the continent. Across this domain of smugglers, immigrants, outlaws, and police, around their walls and down their secret corridors, Berman roams like a holy fool. He does not want to speak Spanish too well. He does not want to explain himself in too much detail to the taciturn men gathered by the cattleguard. Berman smiles and gesticulates. The men say the way ahead is *peligroso*. Berman shrugs a shrug that could mean anything, and the men, laughing, wave him through.

Berman says he feels safe in such places, prowling the back roads in an ancient pickup truck that has already exhausted a cat's worth of lives. He parks it in the loneliest of places, with not a house, not even a power pole between the horizons. He leans briefly under the hood, disconnects something vital, puts the missing part in his pocket, and sets out. On foot he lugs his view camera down washes and up ridges where no trails exist; sometimes he has a plan, sometimes he hasn't. He hunts the angle and moment that shape the visual speech of the land into a clear declarative sentence. The statements he captures are not about ordinary beauty but about extraordinary character. These wordless sentences are the poems he hauls back to his studio and commits to paper. In the fullness of time, if such images do not cure society's indifference to these impressive lands, they will at least memorialize a region that stands just outside the gates of our senses, a region both old and new that scares us a little bit because it looks so raw, as though the creation of the earth still has a long way to go and it doesn't matter whether we make the trip or not.

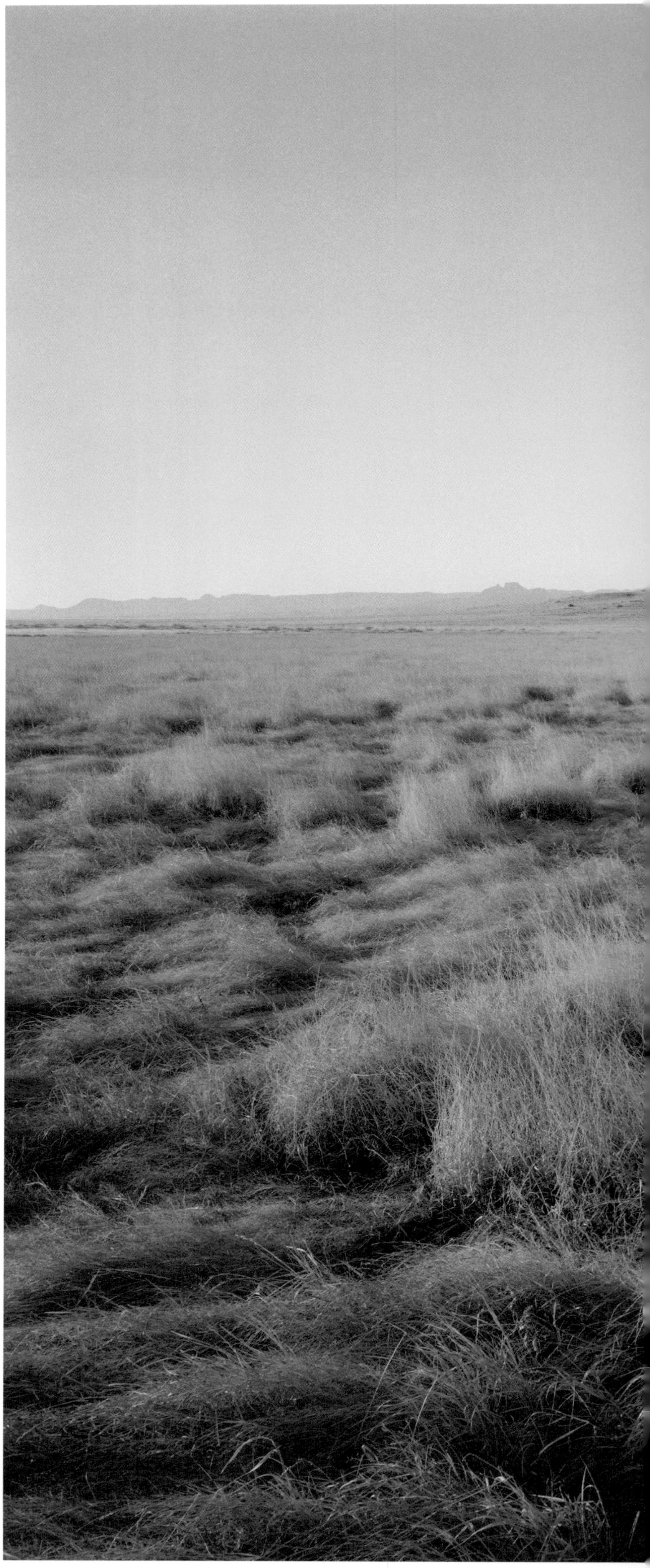

**Rancho Uno**, Janos, Chihuahua

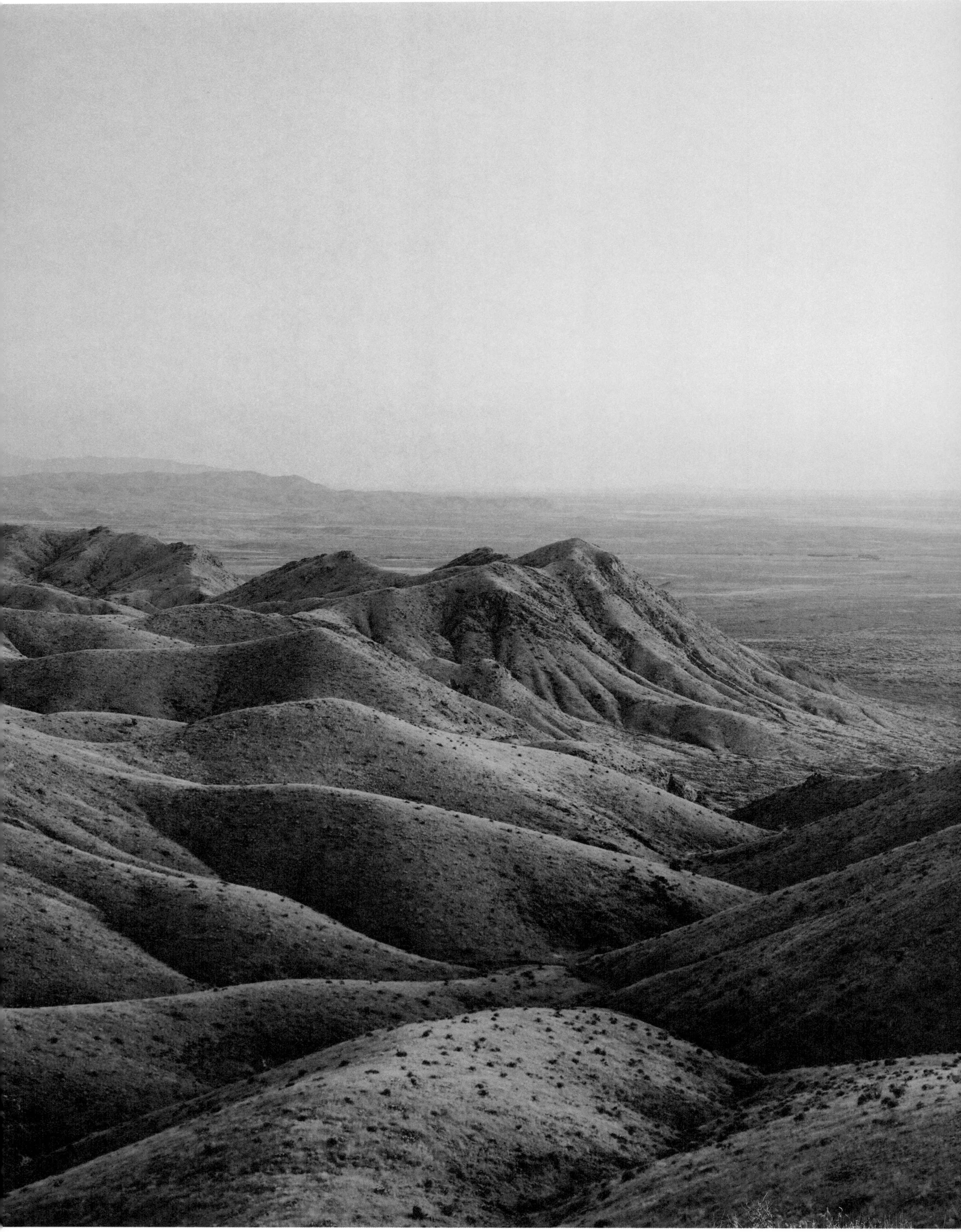

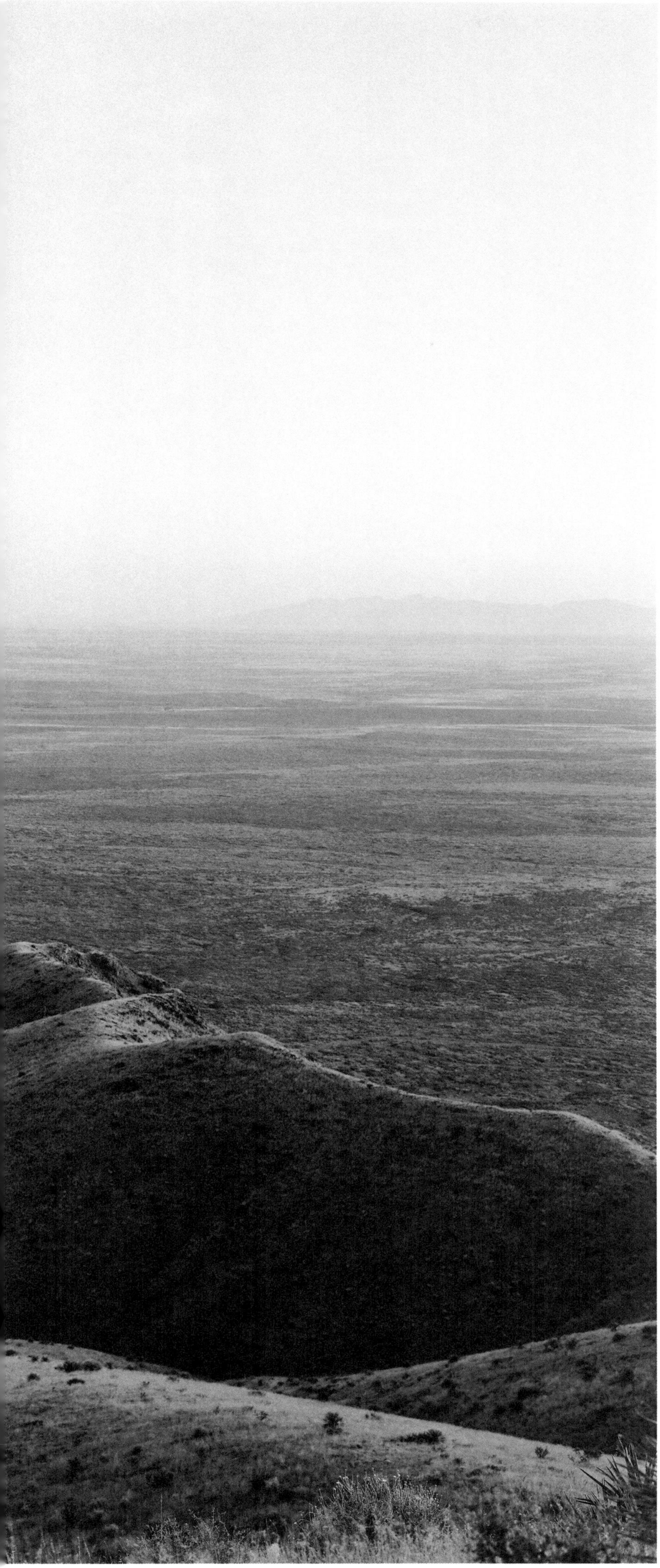

**Godfrey Hills**, New Mexico

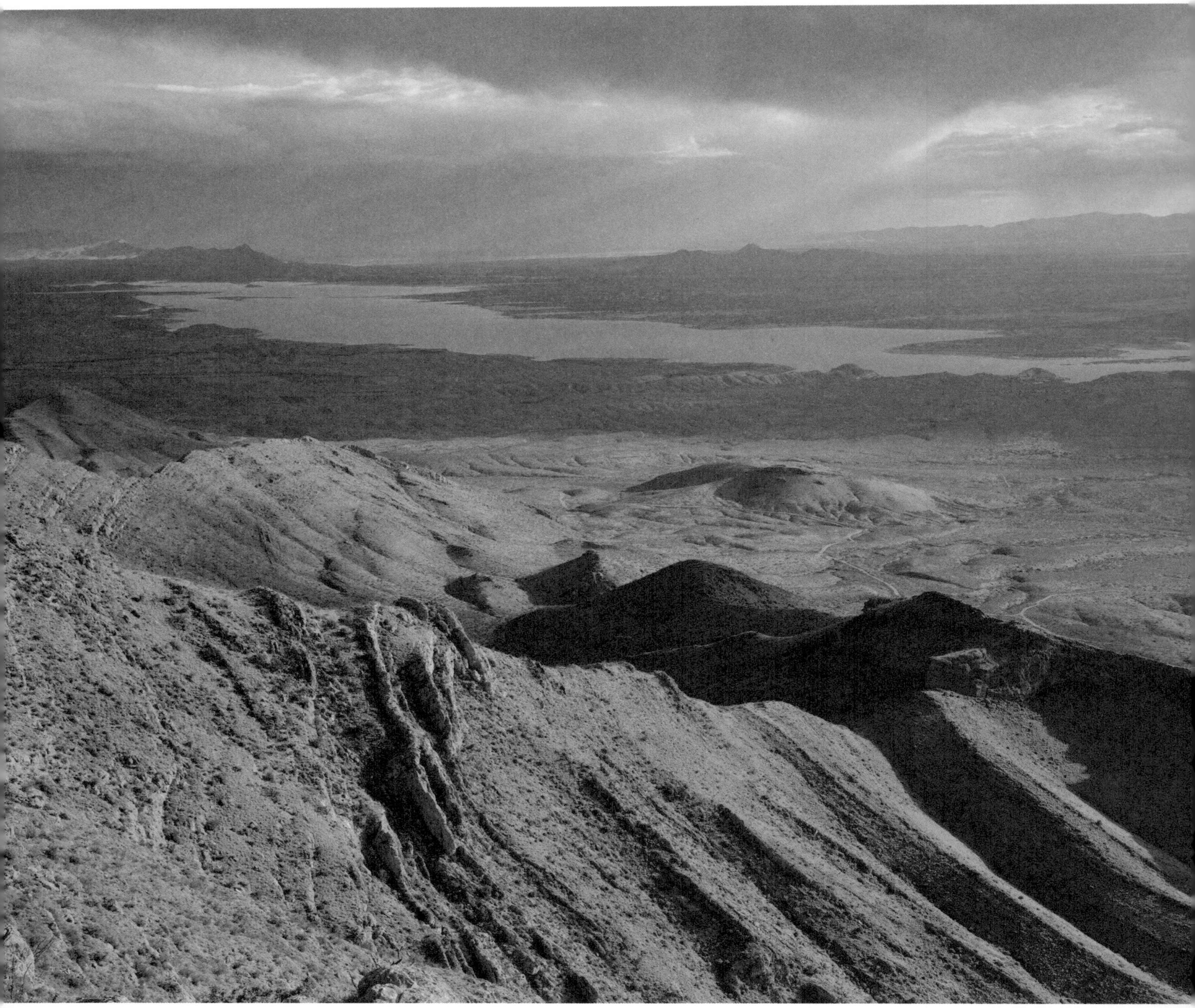

**Fra Crislobal Range**, Armendaris Ranch, New Mexico

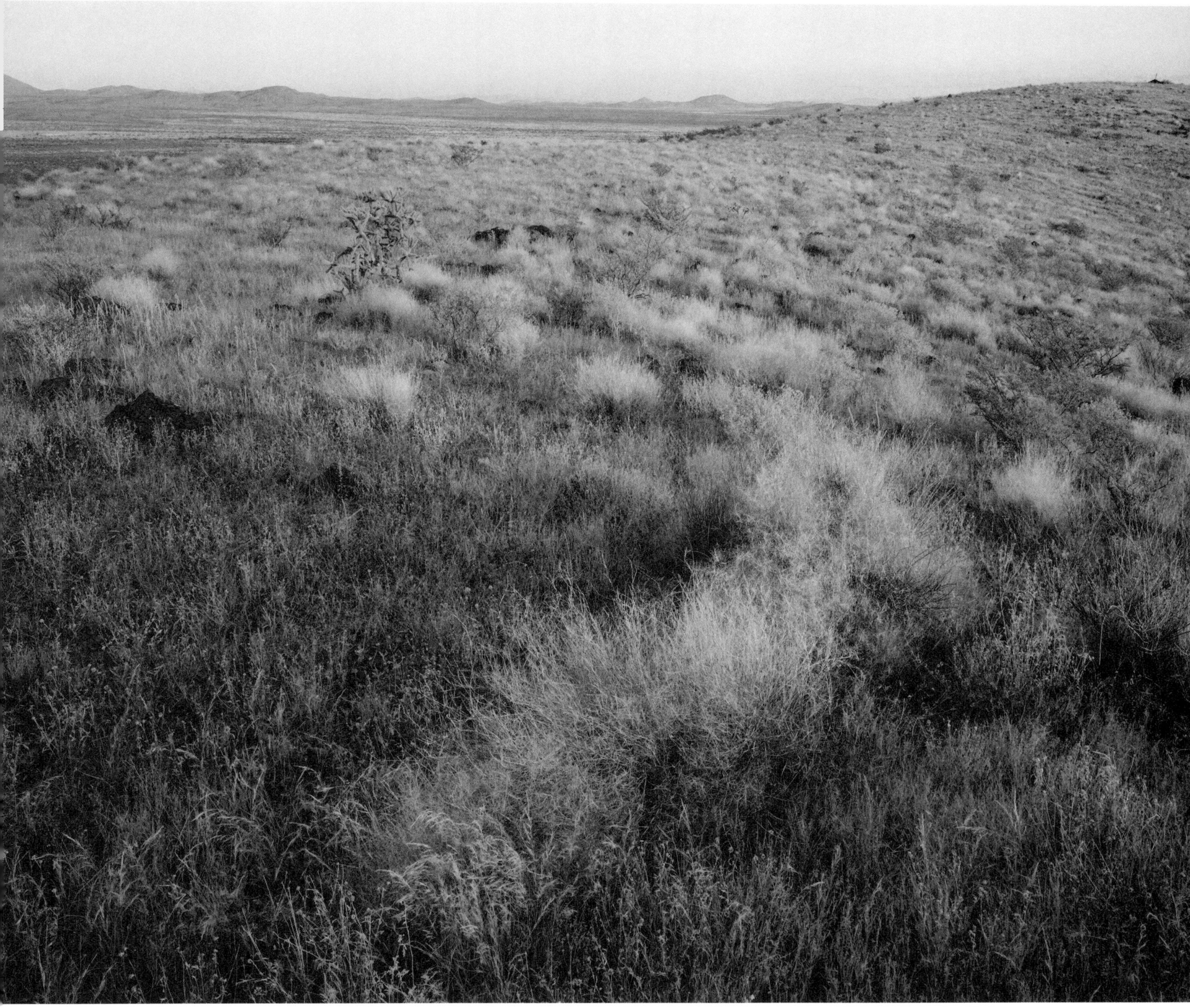

**Potrillo Mountains**, New Mexico

**ELEMENTAL**

BY REBECCA SOLNIT

**THIS WINTER WHEN IT WAS TOO COLD** to go camping anywhere else, friends took me to the deserts of southern Arizona, and then half of them went home. As night fell, we lingerers were left in a silent immensity, four people scattered far apart with silence in-between and all around. In the morning, I watched the dawn come in from my cot, settled between a thorny, delicate palo verde to the west for afternoon shade and an armless saguaro dark against the eastern sky, a soft cloudy dawn and sunrise with clouds tinged gold and moving east after a night that was sometimes starry, sometimes muffled in warming cloud. I arose one morning when the other three were sleeping. I had been away from the desert for a long time and it seemed at first like a tune I'd forgotten how to dance to, a place I couldn't enter, though I could look at it, but this time I began to catch the rhythm, which was of many kinds of time all beaten together, and the solitude, and the slowness of immersion in the immensities, a wealth made of many kinds of lack and absence.

There had been days and nights of not having to separate myself from the world, even enough to get into a car or a tent, let alone a building, not looking through glass or into mirrors and reflections and barriers, but fully immersed in an undivided world; a world that was harsh and threatening in some ways, but in others enormously seductive, something you could melt into, though it always seemed to promise that to melt into it all the way would be to become a pillar of salt like Lot's wife, or a pile of bleached ribs in a nameless wash, or just the wind scouring the peaks promised the mortality that is never very far away in the deep desert. I remind myself again and again that if deer died out here, then they were also born, and so were birds, lizards, trees, shrubs, grasses, all those things that one seems to come across most often after they have dried up and stopped moving, as though life itself were elusive to survive these elemental forces. And if death is not, then it is, because here it is not hidden even if it is no closer than in the humid and crowded realms.

The place we were had no name on the maps and, beyond the road, few traces of people—some shotgun casings, a light on the far horizon some dozens of miles away—it was not one of the celebrated, managed parts of the desert, just part of the expanse outside the bustle most of us inhabit, a remainder in some sense of the dreamtime. I got up in the early light and wandered east in a straight line from the cot on that south-facing promontory, past little hills of pale volcanic stone and red hills, and hills with a lovely bright flush of green on them in this moist desert winter. The morning before I'd gone only slightly to the north and come to a big cliff overlooking the canyon that opened up to the endless valley, or basin, to the south, but this time I came to a slope that led down into it via little washes with the occasional small remnant pool of water that even stagnant and gnat-ridden looked like a blessed thing. The washes led me down into the lusher lowland where we'd found desert lavender and much else in bloom on the first day. The place smelled wonderful, the smell of the desert I'd forgotten, creosote's fine familiar aroma everywhere and occasional wafts of sweetness, pockets of warm and cool air, and the sharpness of the

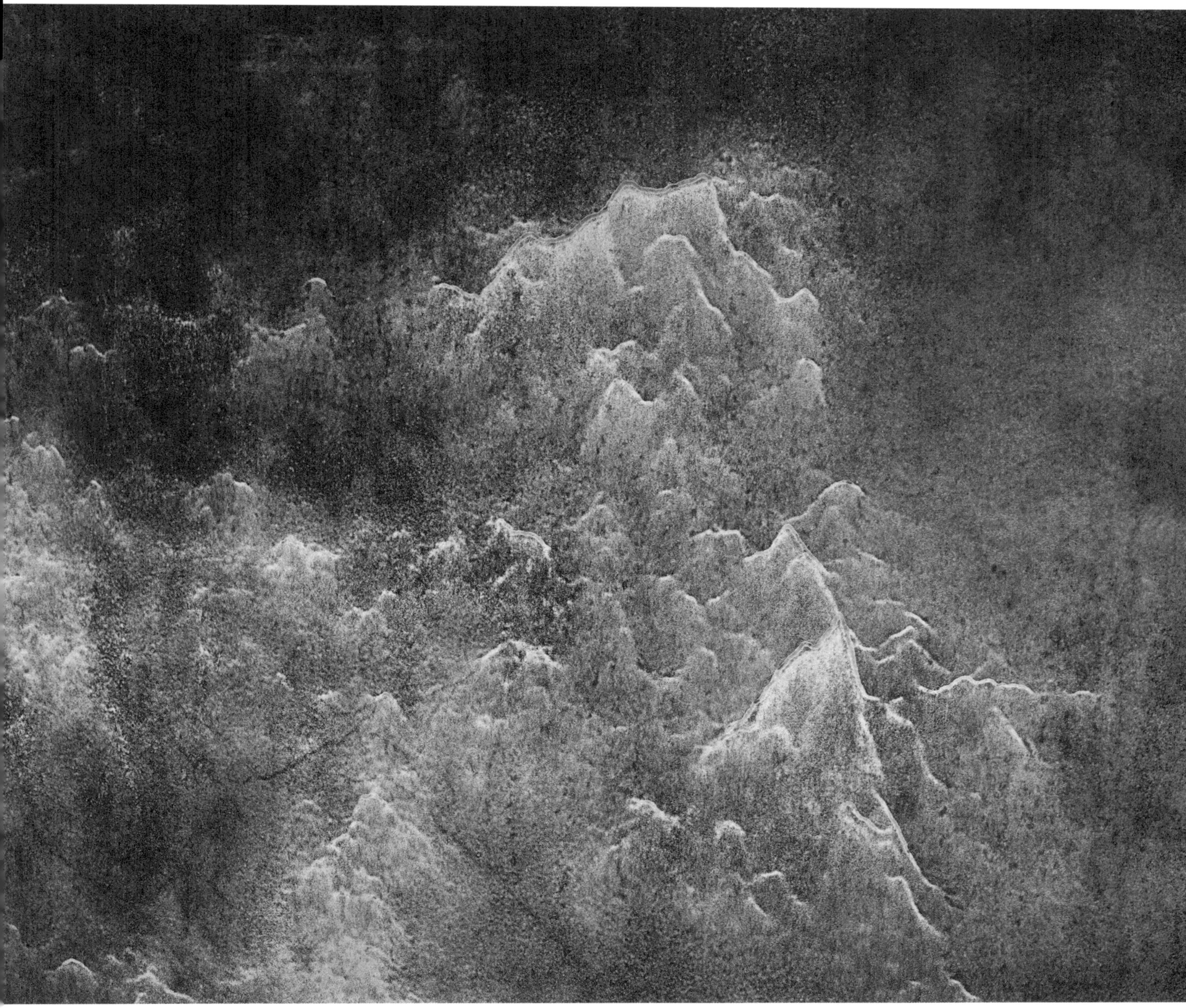

**Salt, Steel and Tar**, Cornudas Mountains, New Mexico

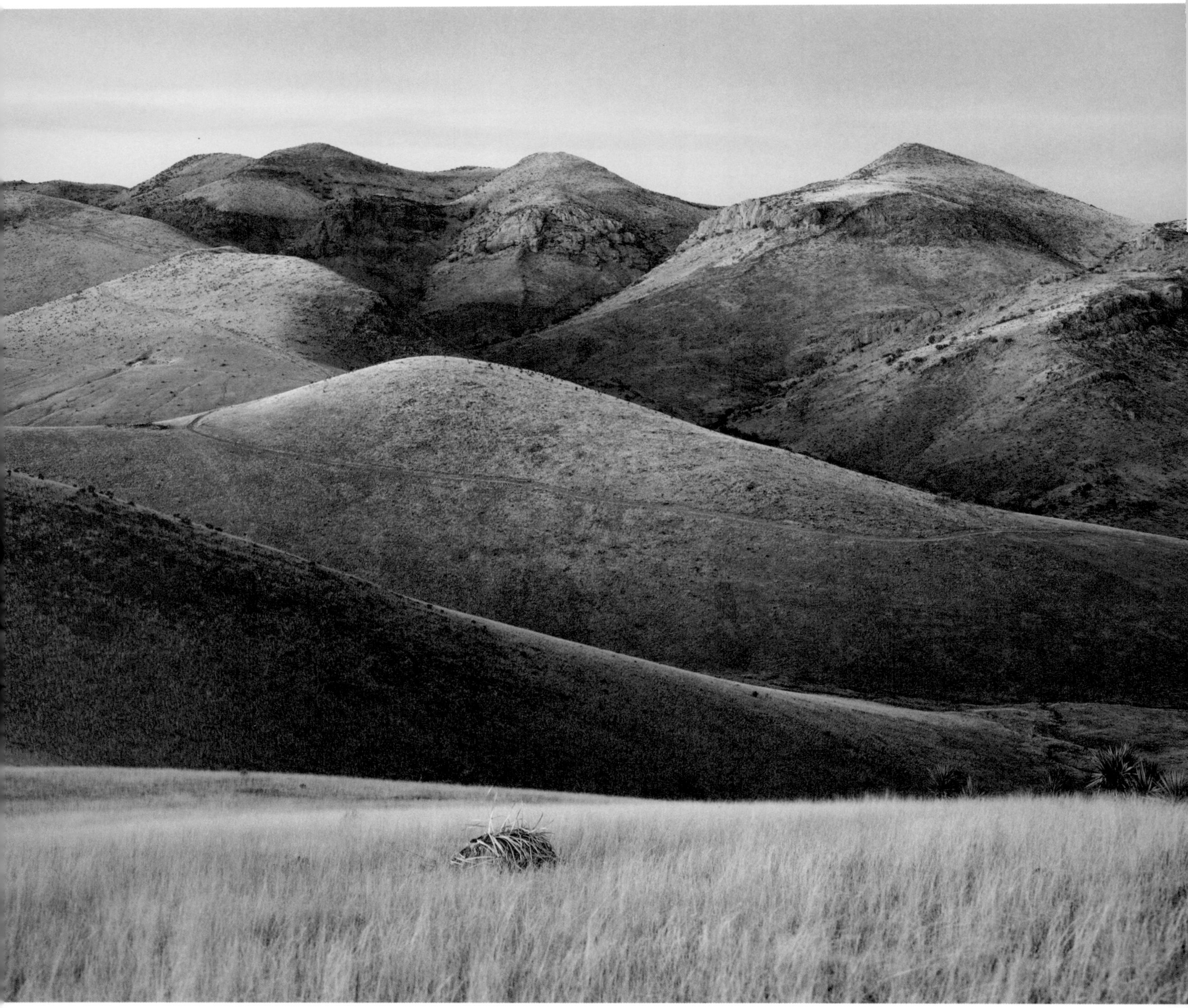

**Sierra el Capulin**, Rancho Corralitos, Chihuahua

lavender and on it the hum of bees. It was the kind of place where you have to always watch your step, uneven loose stones and cholla—which New Mexicans call horse crippler—everywhere waiting to stick to you; pulling a spiny chunk of cholla off someone else's ankle is one of the routine kindnesses on our expeditions in the Sonoran.

So much of this landscape for me is the tiny scale of insects and flowers and pebbles and then looking up to see the rugged formations and the blue of distance on the horizon, as though some kind of familiar middle scale were absent, but like mortality this too may be the true scale of things made more evident in this clear laboratory of light. That morning I had to try hard to not get lost, though getting lost had its own allure, an immensity one wants to be swallowed up by, a promise that this all went on forever, even if I couldn't.

Instead I curved around to the big basin, though down in the washes the vastness wasn't so visible. I traveled through a wash that was red and then lavender and then, as I went uphill to a stone arch framing it, through the sometimes waterway of pale stone with dark pits where the water spent the most time. From there, on this day full of landscape and empty of incident, I managed to scramble up a lot of steep slope and streambed to arrive back at a camp that was still uninhabited.

Grass grows and dries and dies, clouds drag their shadows across a rough plateau, a goat dies on a fence, a lizard crawls across a dune, a yucca finds enough water to grow as tall as you, light shifts, and in the longer scale of hundreds of thousands of days and nights chasing each other around and around again, rivers carve canyons into the landscape into which volcanoes extrude cinder cones. In the deep desert, many of the landforms have no names, the dead whose flesh was devoured or dried to leave bleaching bones under the burning sky have few mourners; the changes have few witnesses. The continued existence of places that are wilderness not in the sense of the pristine, but of the profound and remote, remote not just from towns and paved roads, but far from the surveillance and management that has come to epitomize our time is grounds for relief and even for joy—a somber joy that knows what we have to gain from knowing solitude, mortality, difficulty, distance, and slowness. The border matters here, and climate change, and economics, and the belief systems that on the one hand leave small shrines across the desert, and on the other impose cattle on it; but this place is still beyond much of what most of us are immersed in every day. The pictures here are news that there are still places beyond the reach of the news, places that are not much part of the reality most news and entertainment describes, and those who make these stories that are about the bustle inside the phenomena called society and economy hardly comprehend that such places exist still. But they do.

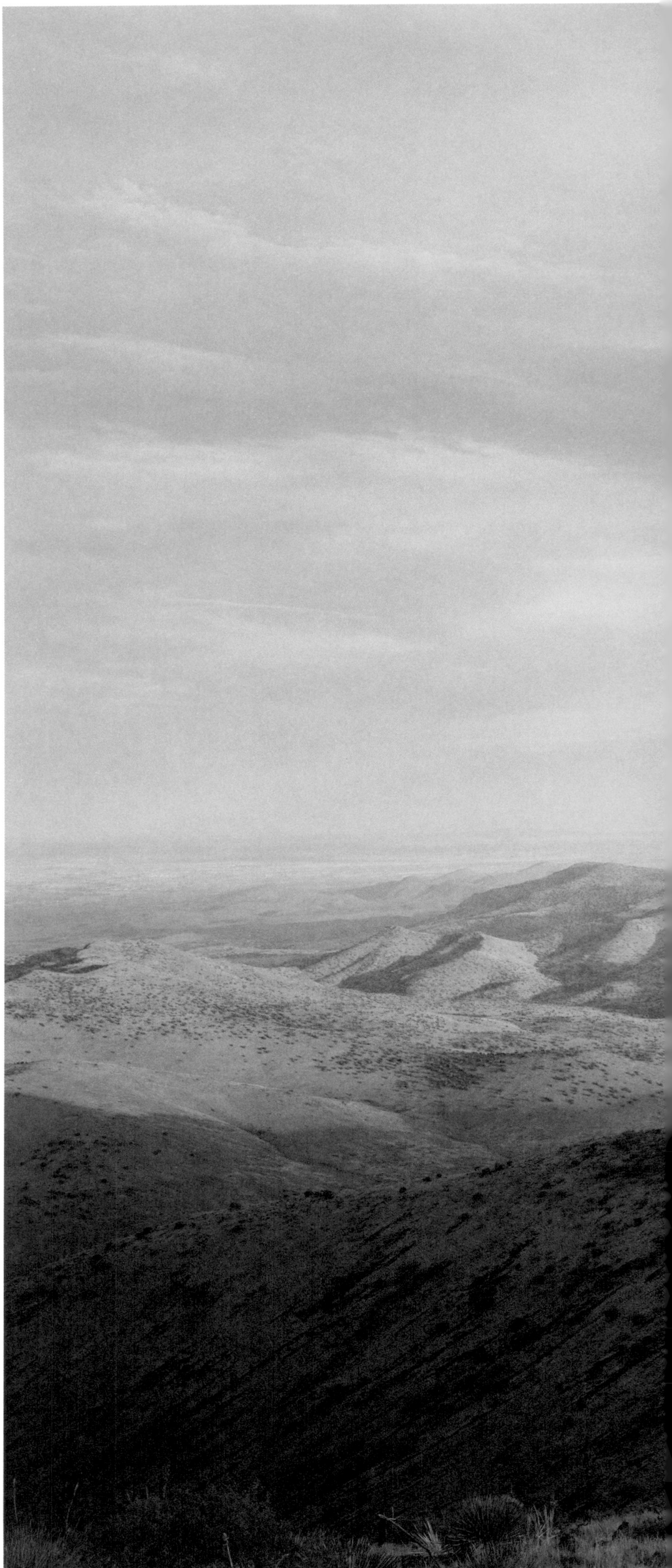

**Peloncillo Mountains**, Arizona

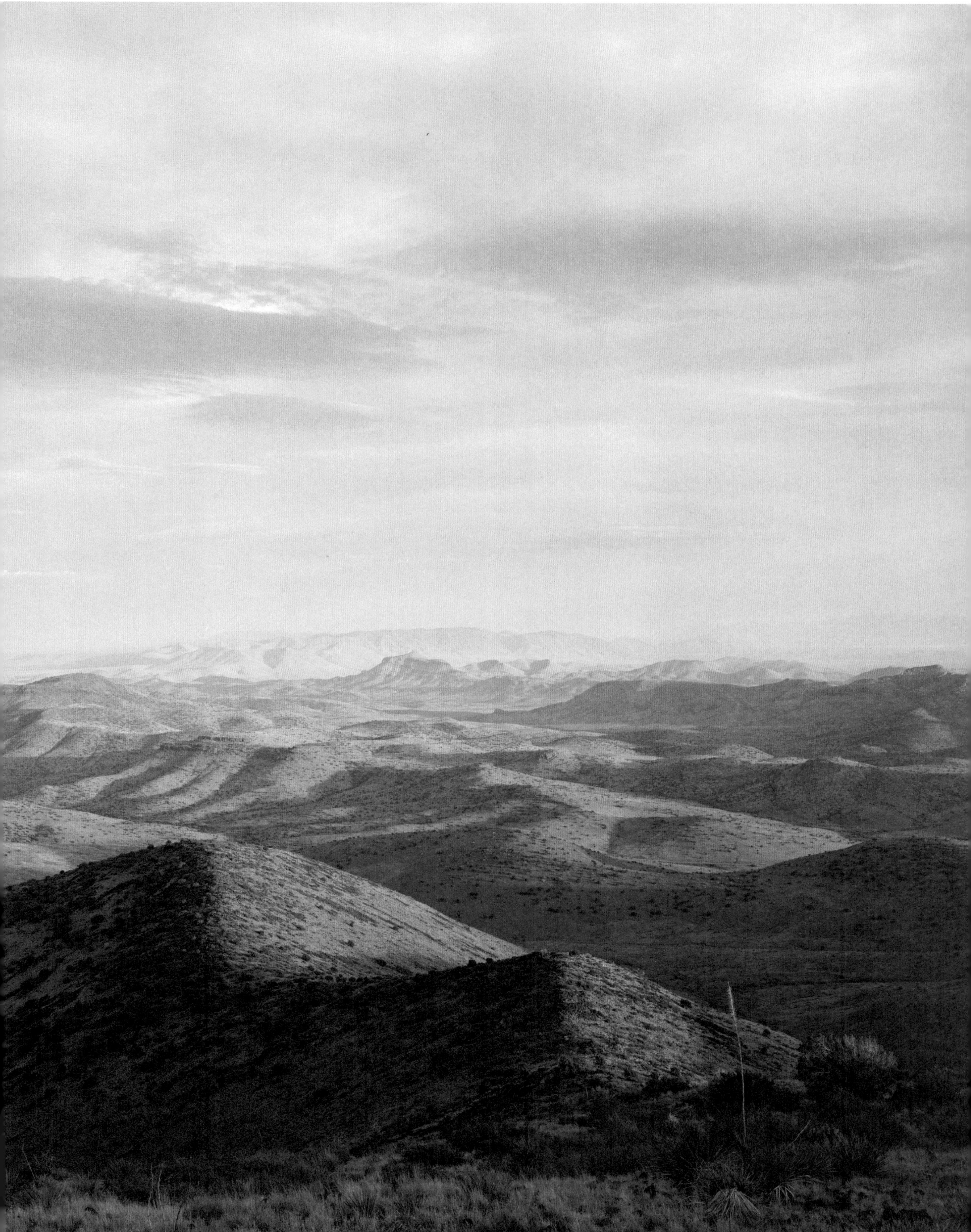

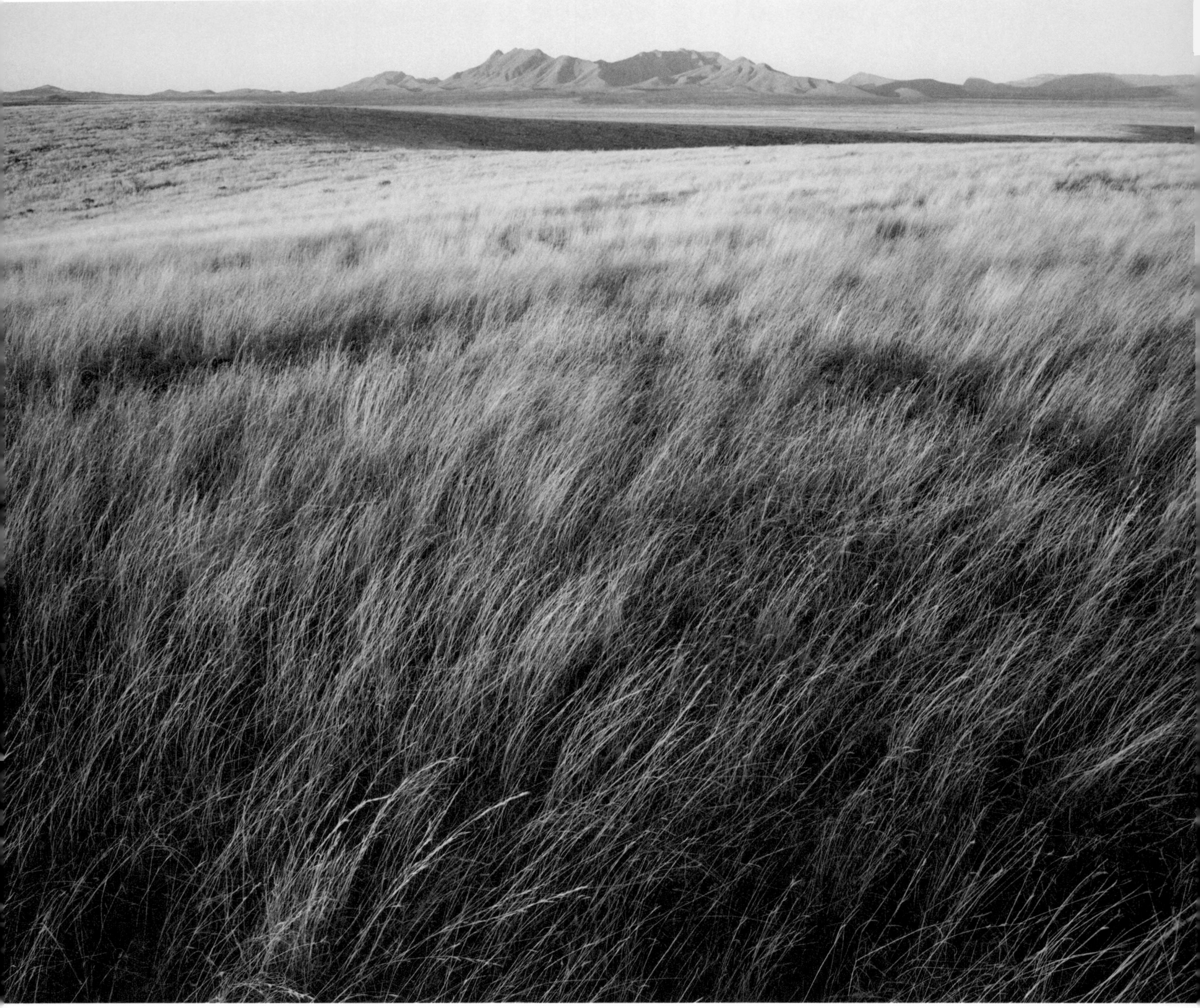

**Grasslands**, Rancho Tinaja Verde, Chihuahua

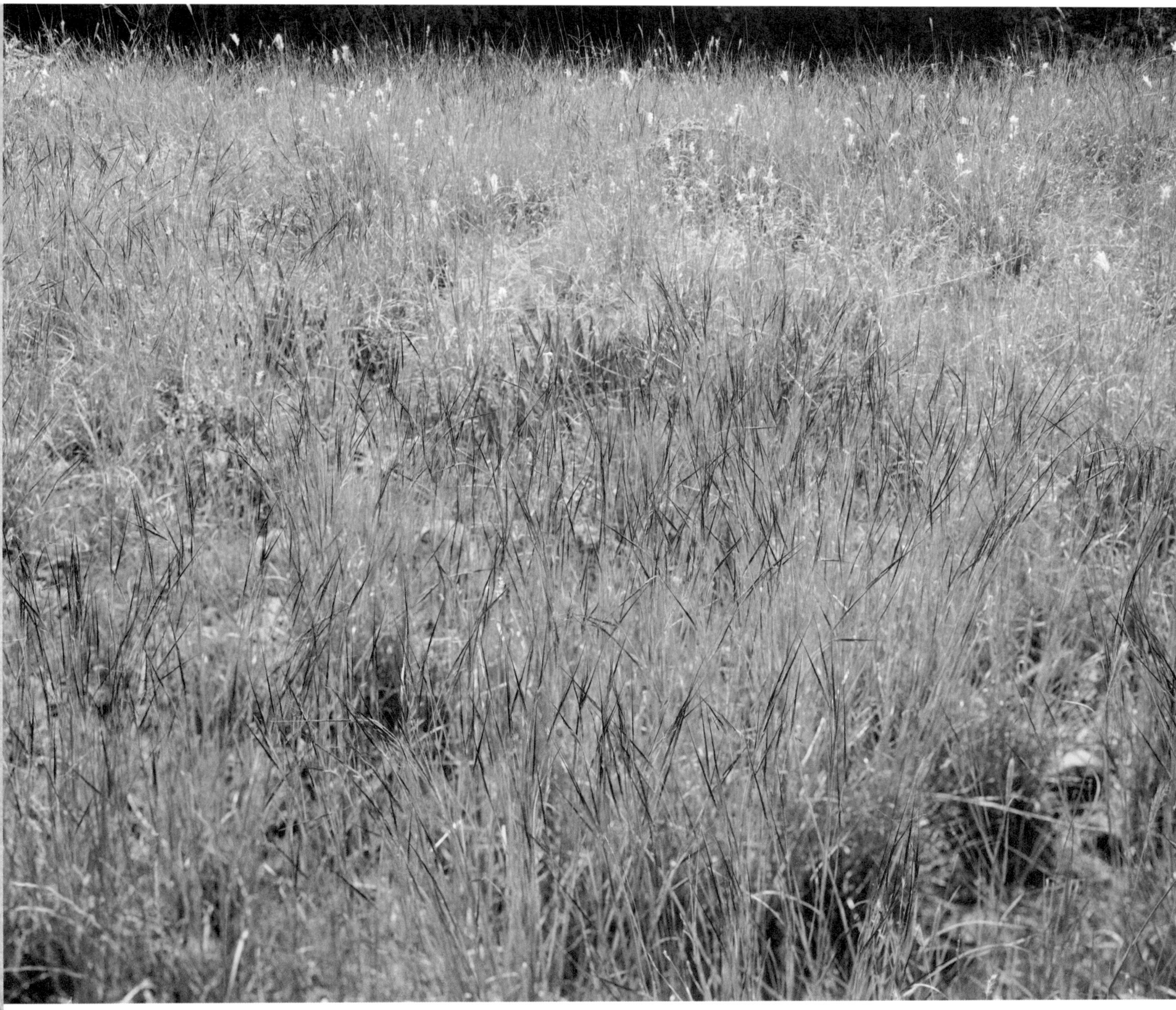

**Tanglehead**, Sierra Vieja, Texas

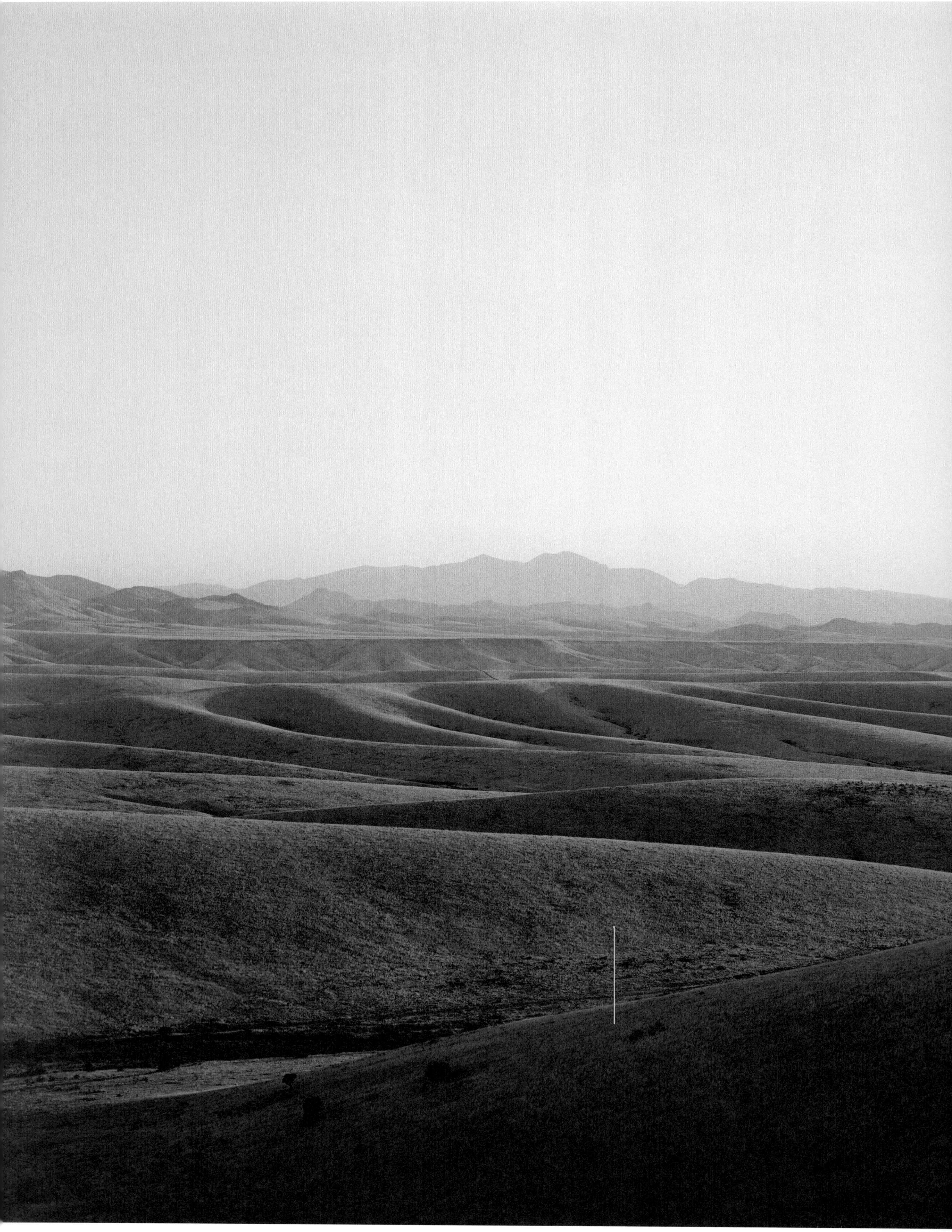

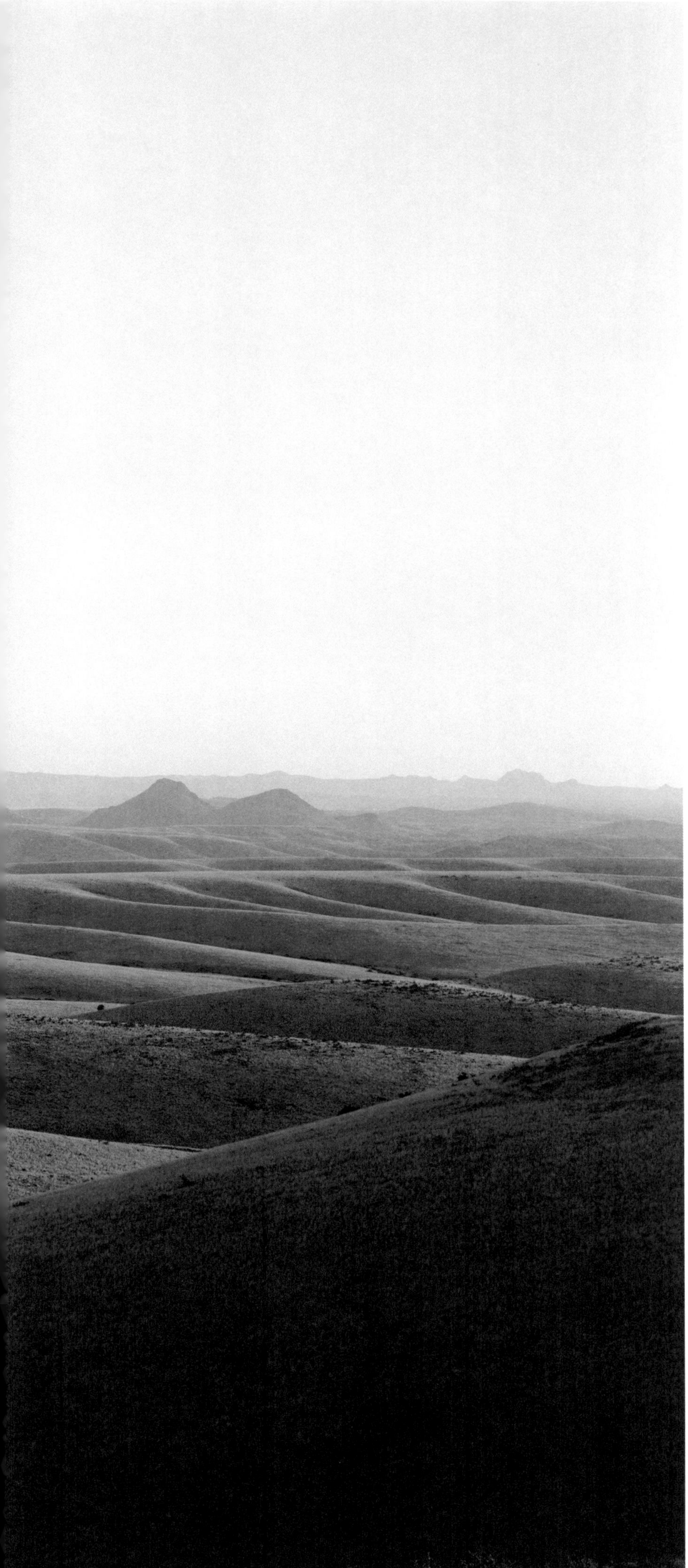

**Antelope Canyon**, Ladder Ranch, New Mexico

This first printing of *Grasslands/Separating Species*
is limited to 1,000 copies.

**RADIUS BOOKS** is a tax-exempt 501 (c) 3 nonprofit organization, founded in 2007, whose mission is to encourage, promote, and publish books of artistic and cultural value.

**RADIUS BOOKS**
1012 Marquez Place #109B, Santa Fe, NM 87505
www.radiusbooks.org
Design: Skolkin + Chickey

**516 ARTS**
516 Central Avenue SW, Albuquerque, NM 87102
www.516arts.org

ISBN 978-1-934435-21-2
Library of Congress Cataloguing-in-Publication Data available from the publisher upon request.

Printed and bound in Malaysia.

Cover image: detail of *Grasslands, Rancho Tinaja Verde, Chihuahua*

This first printing of *Grasslands/Separating Species*
is limited to 1,000 copies.

**RADIUS BOOKS** is a tax-exempt 501 (c) 3 nonprofit organization, founded in 2007, whose mission is to encourage, promote, and publish books of artistic and cultural value.

**RADIUS BOOKS**
1012 Marquez Place #109B, Santa Fe, NM 87505
www.radiusbooks.org
Design: Skolkin + Chickey

**516 ARTS**
516 Central Avenue SW, Albuquerque, NM 87102
www.516arts.org

ISBN 978-1-934435-21-2
Library of Congress Cataloguing-in-Publication Data available from the publisher upon request.

Printed and bound in Malaysia.

Cover image: detail of *Snow Geese Flight* by Krista Erlick

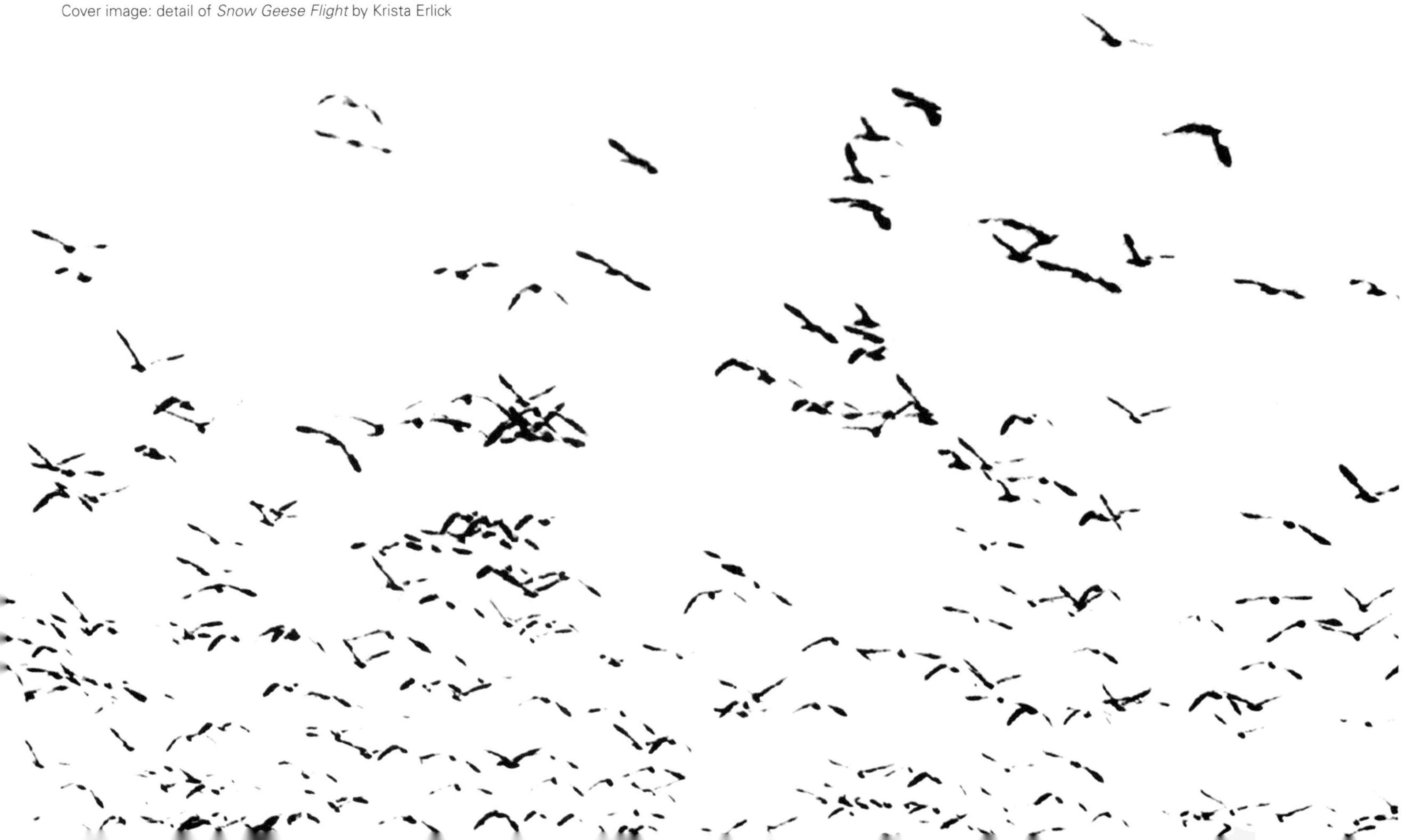

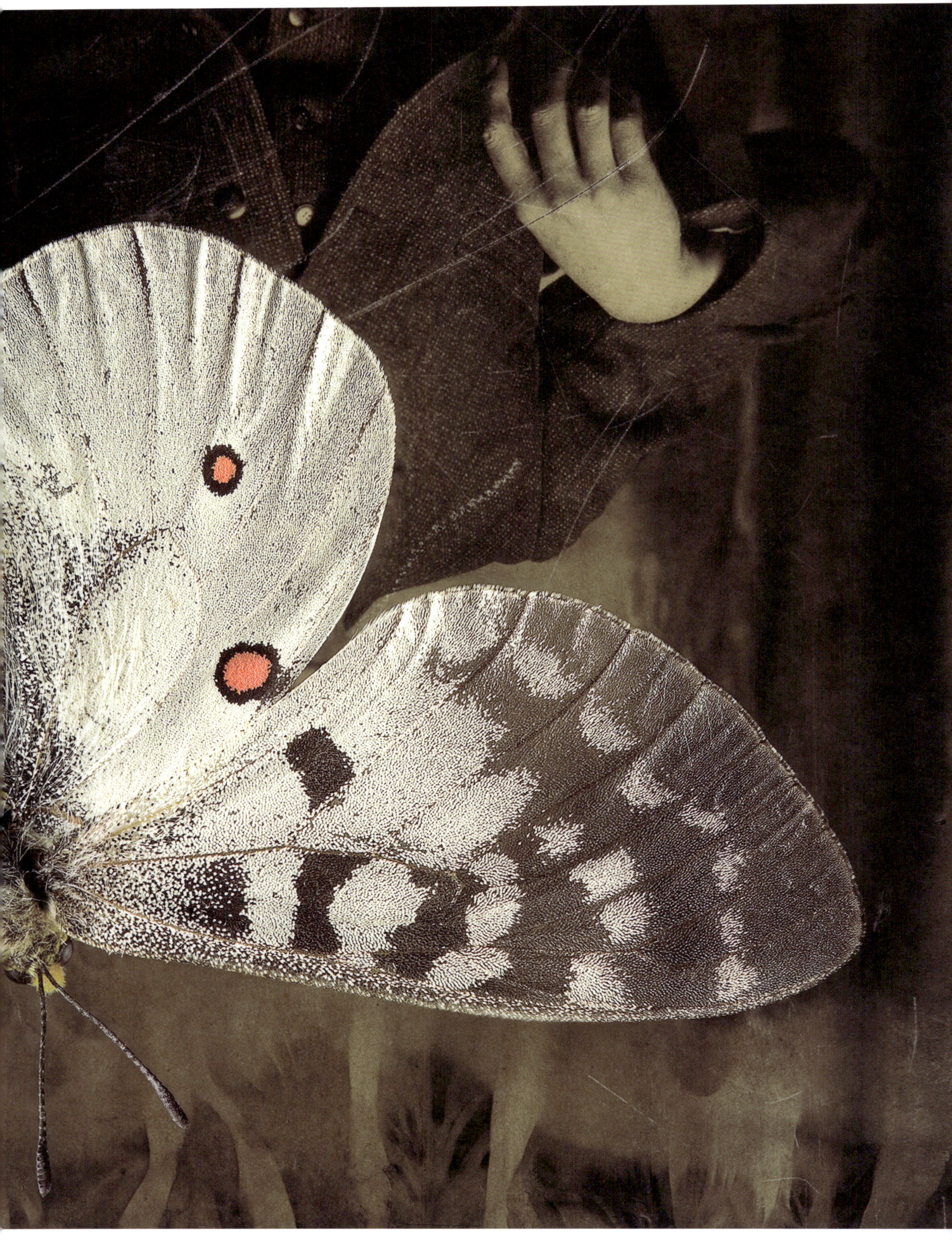

**JO WHALEY**

***Parnassius apollo*** **2007**
Archival pigment print, 24 × 20 inches
Courtesy of the artist and photoeye Gallery

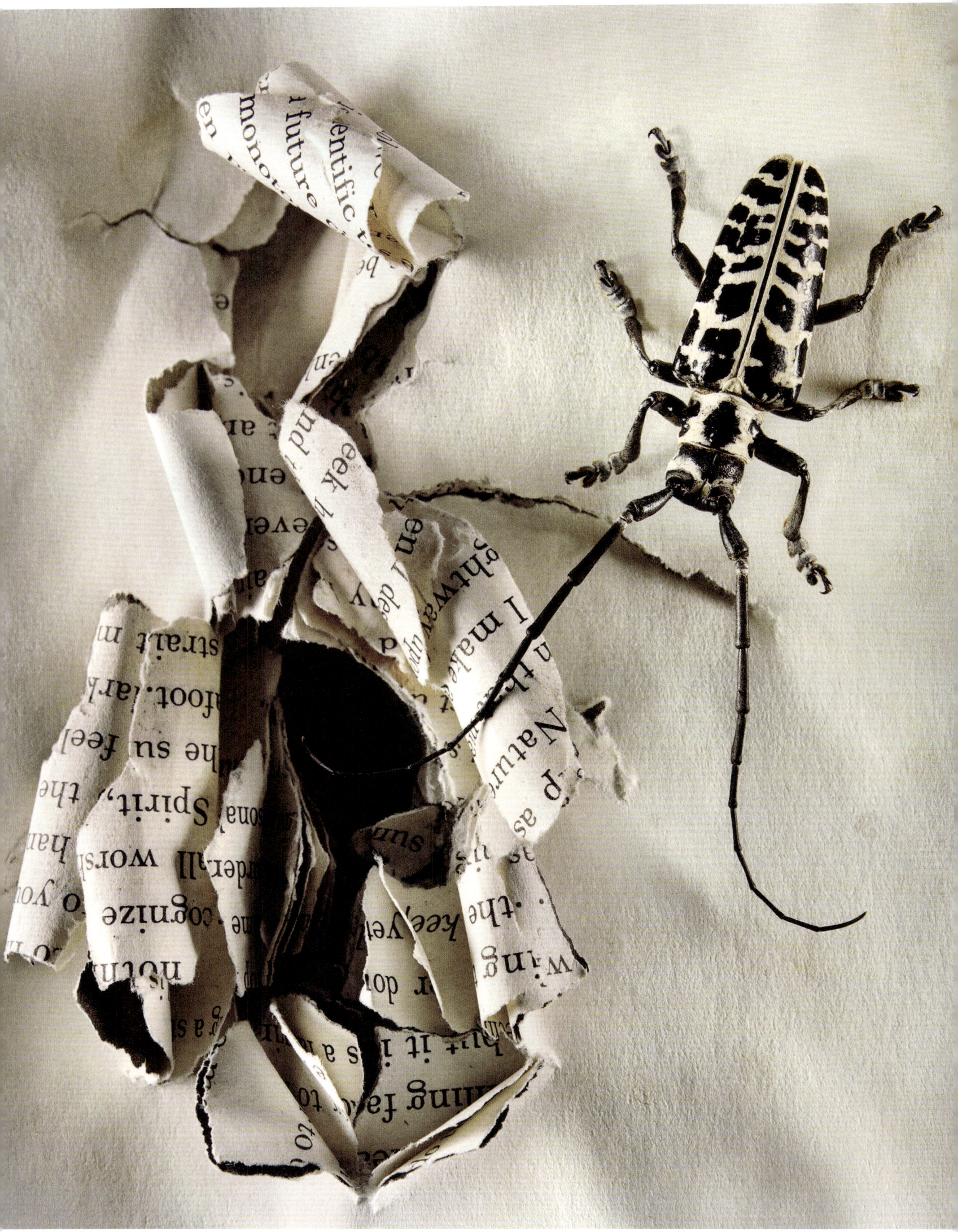

future
I make
Nature
Spirit,
the key,
but it is a

**JO WHALEY**

***Plectrodera scalator*** **2006**
Archival pigment print, 24 × 20 inches
Courtesy of the artist and photoeye Gallery

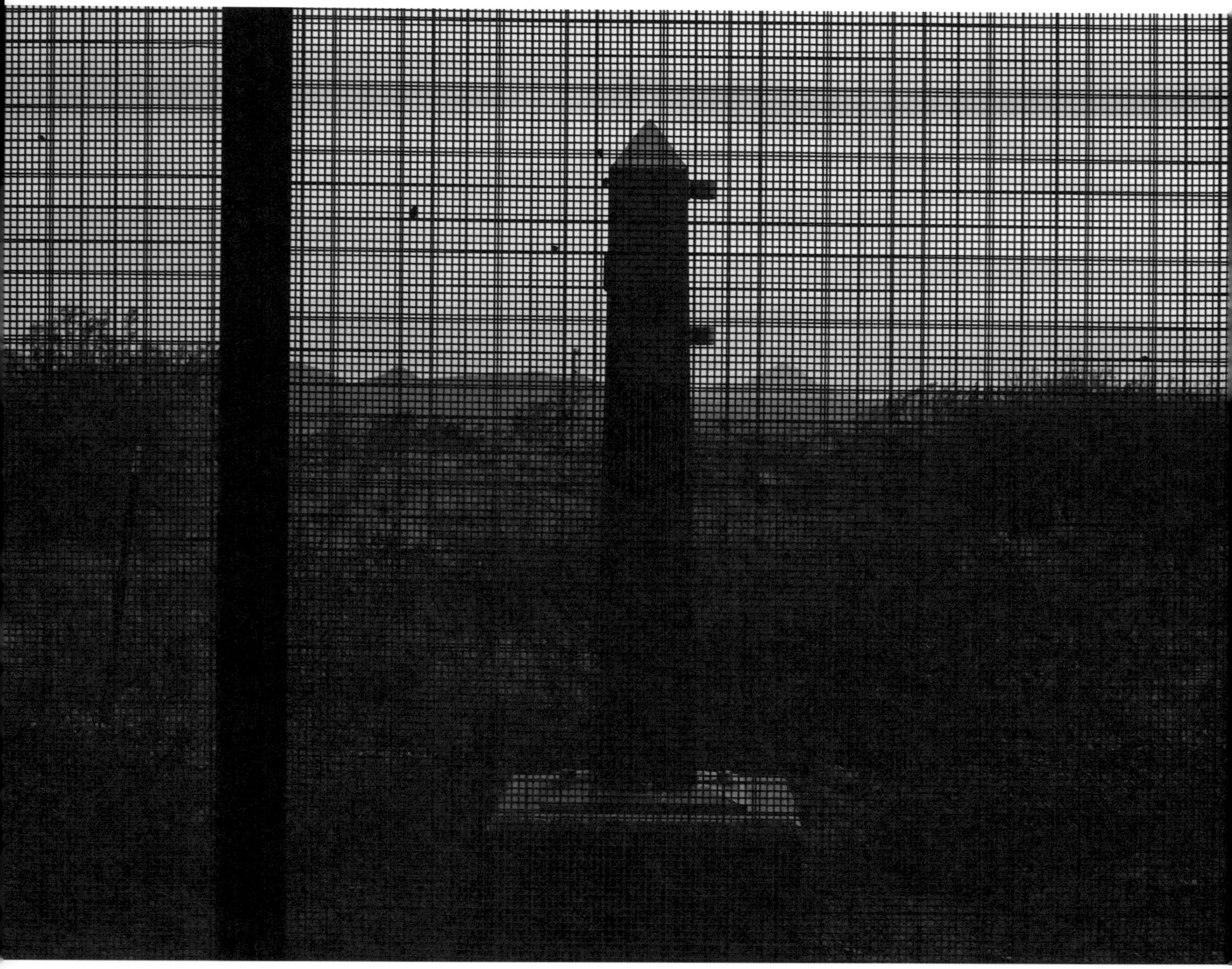

***Border Monument 90*** **2008**
Archival inkjet print, 20 × 26 inches
Courtesy of the artist

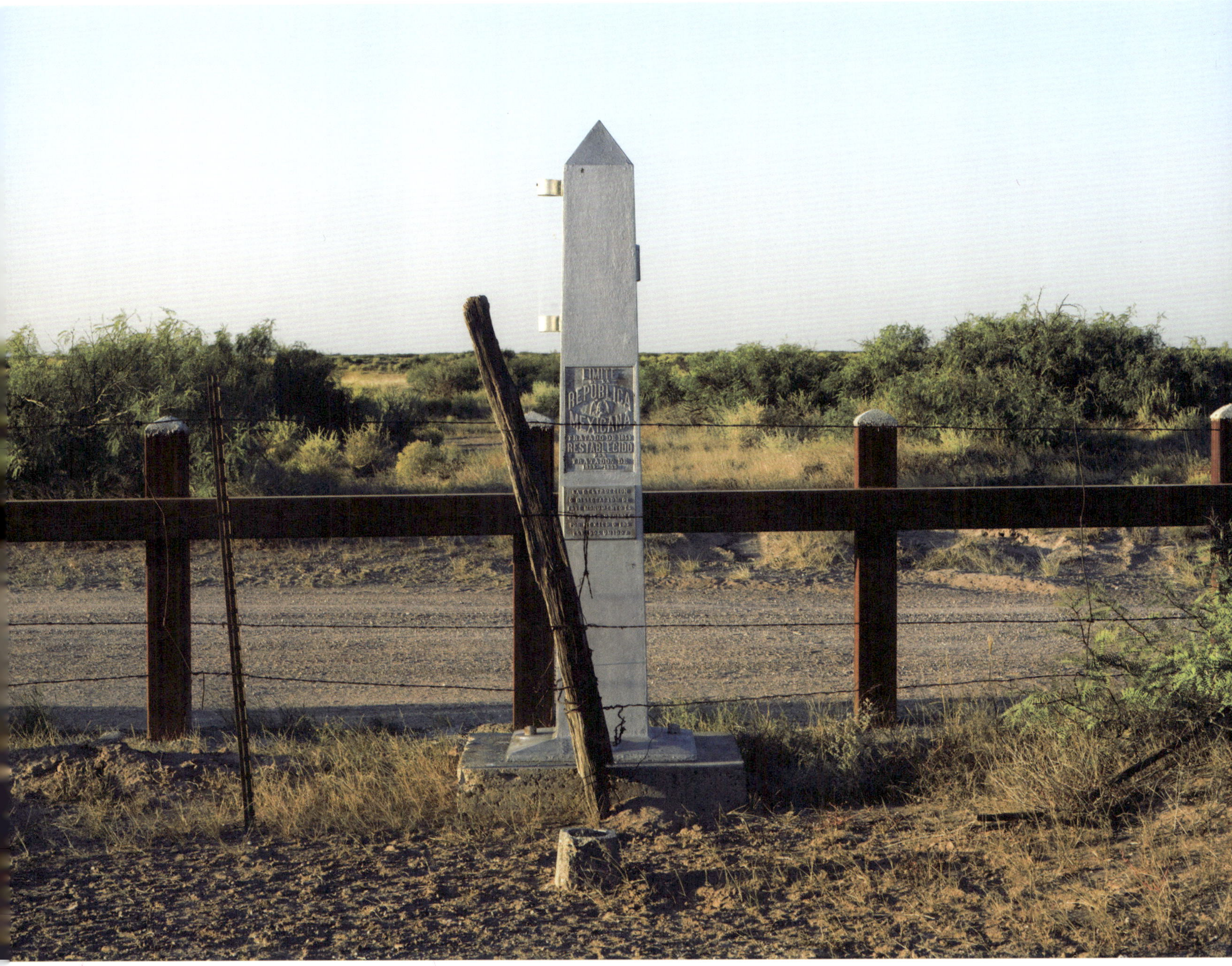


**DAVID TAYLOR**

***Border Monument 18*** **2008**
Archival inkjet print, 20 × 26 inches
Courtesy of the artist

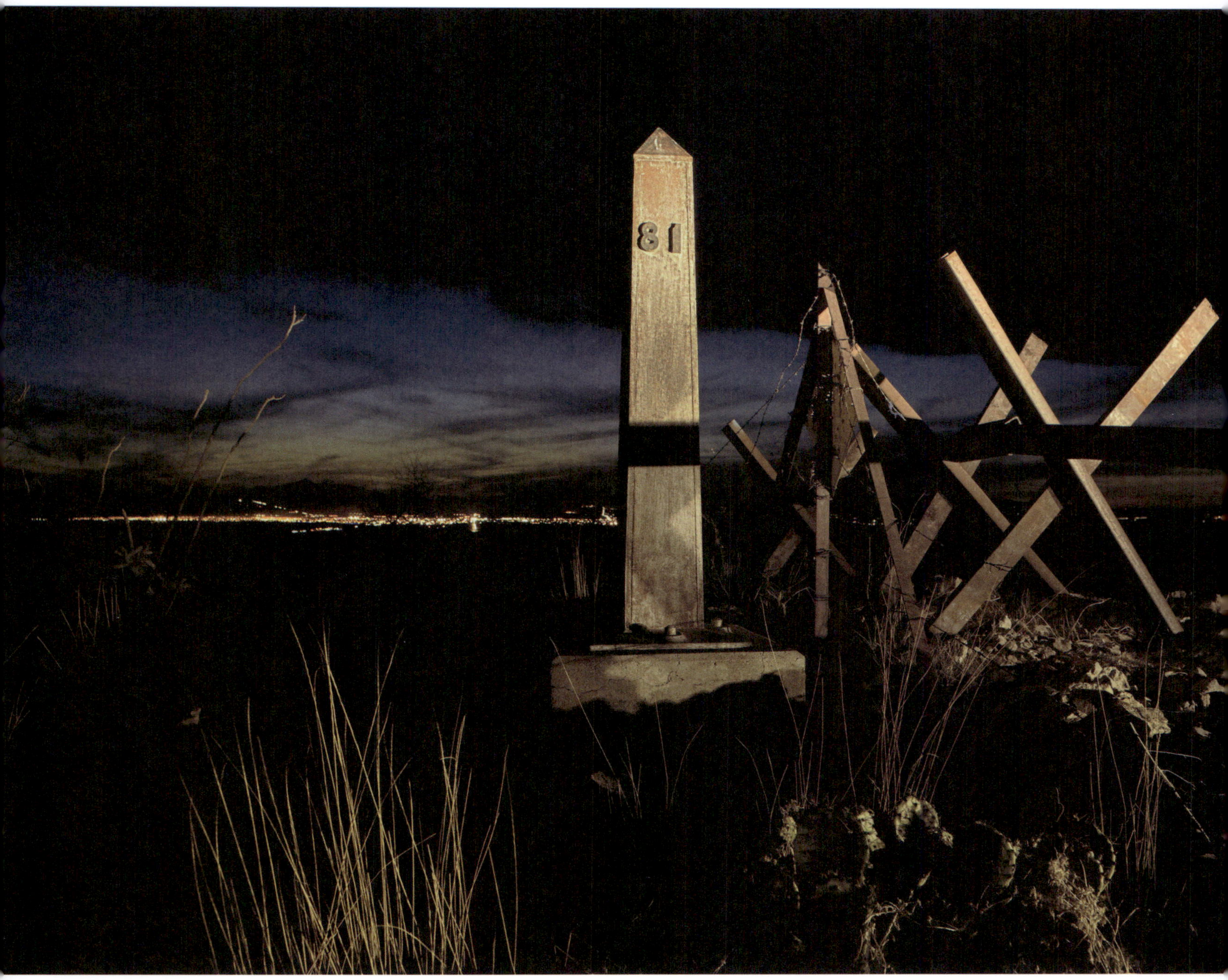


***Border Monument 81*** **2008**
Archival inkjet print, 20 × 26 inches
Courtesy of the artist

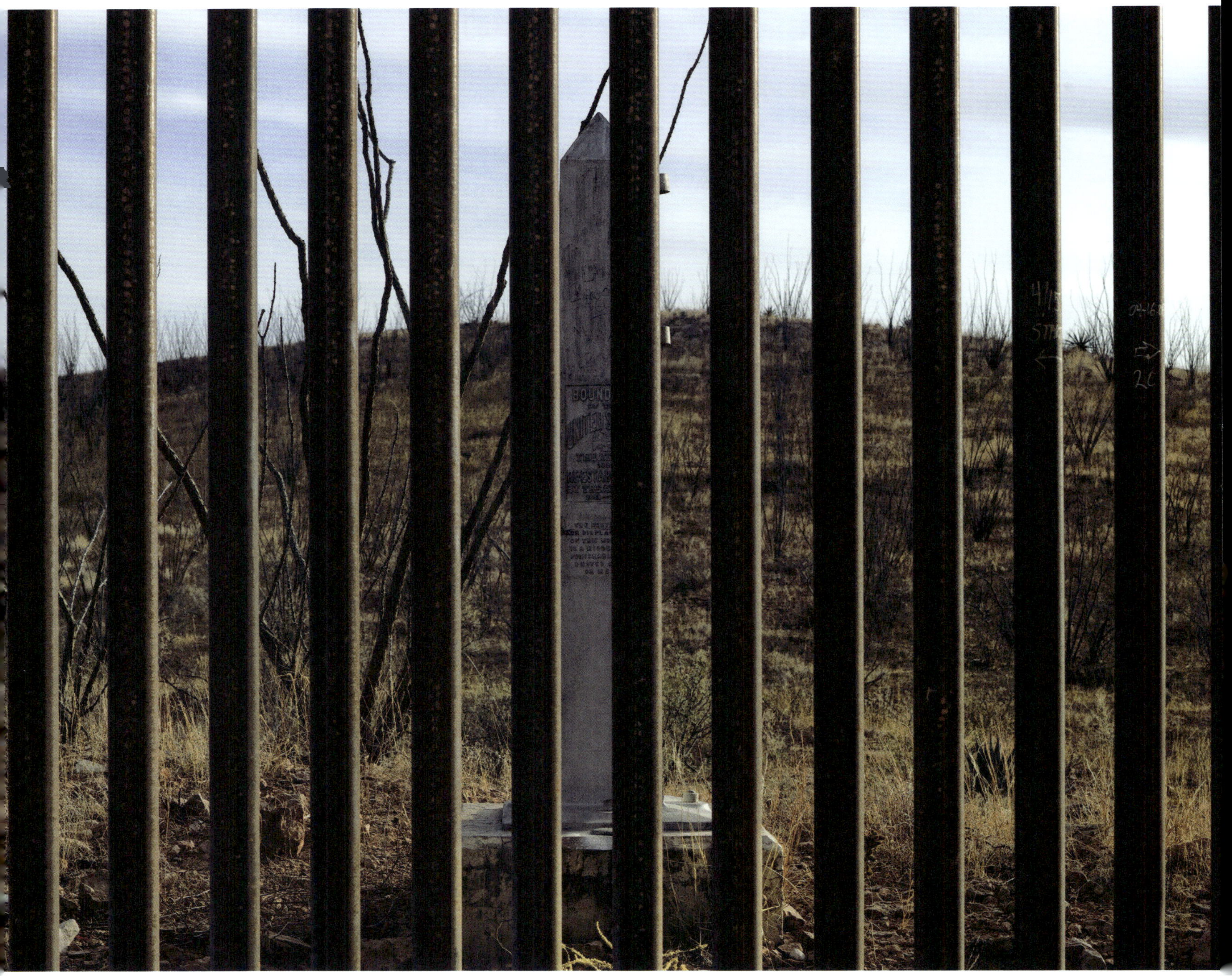

**DAVID TAYLOR**

***Border Monument 89* 2008**
Archival inkjet print, 20 × 26 inches
Courtesy of the artist

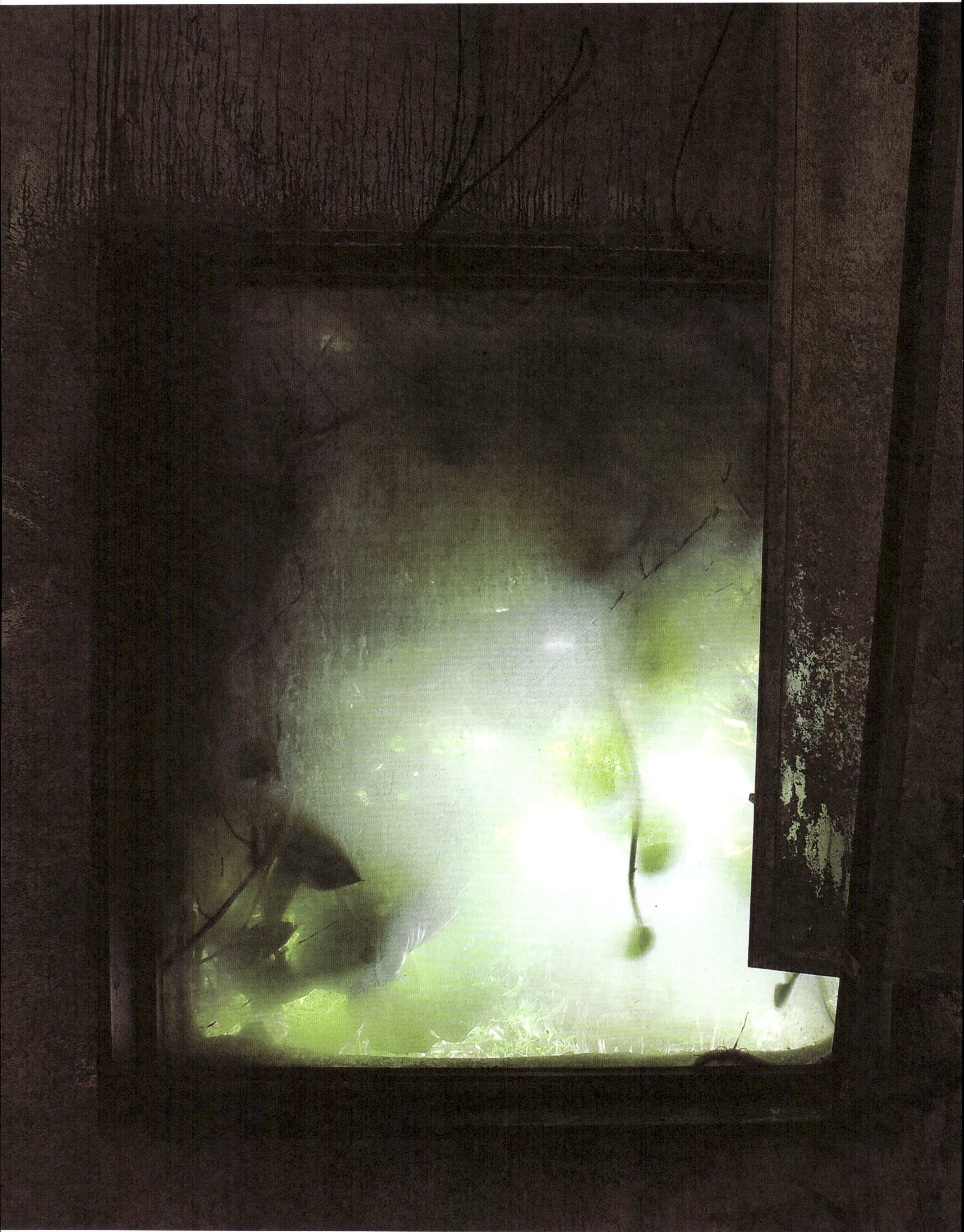

**DANA FRITZ**

***Rain Forest Back Room, Biosphere 2*** **2007**
Archival inkjet print, 12 × 18 inches
Courtesy of the artist

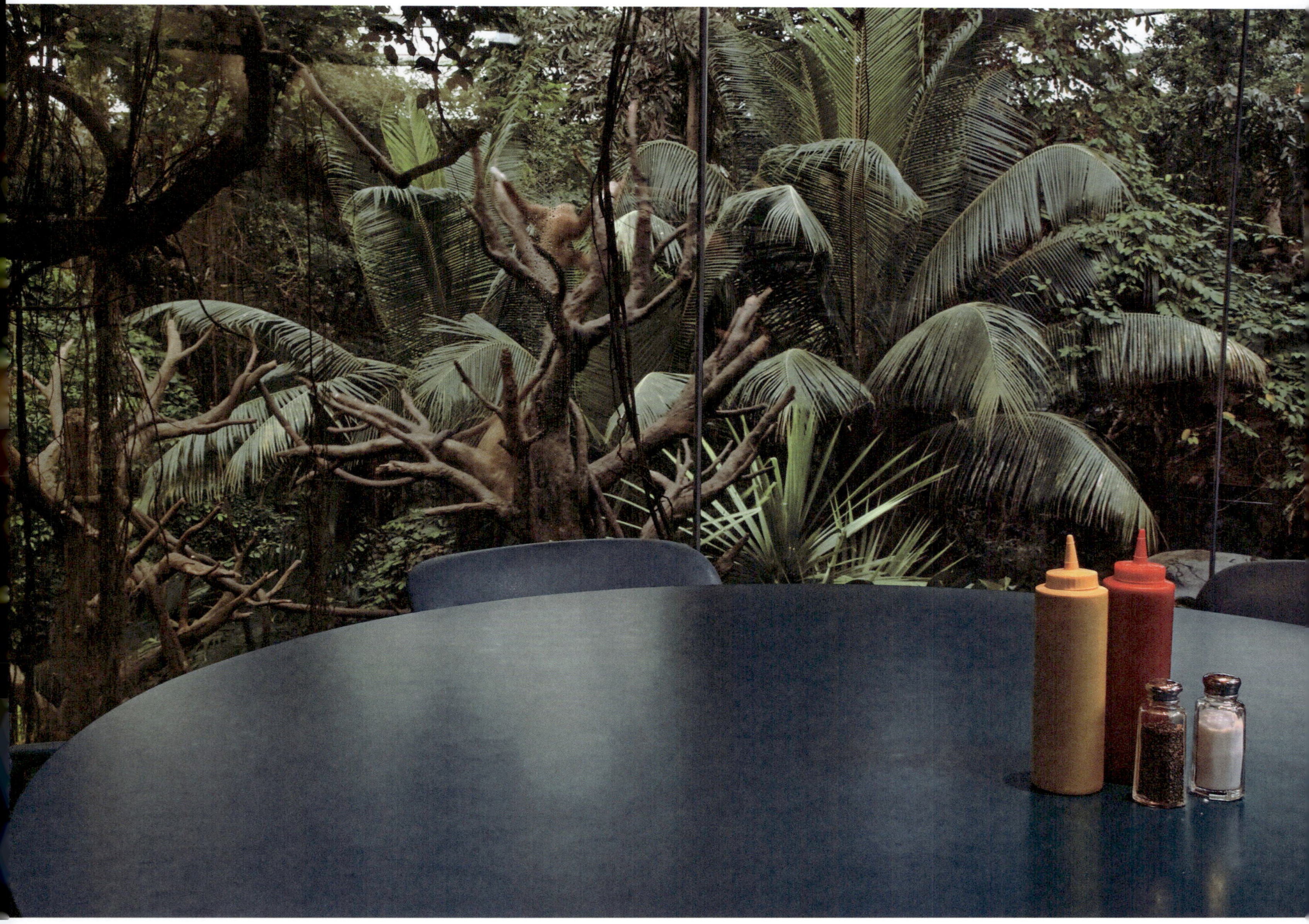

***Condiments, Lied Jungle*** **2007**
Archival inkjet print, 12 × 18 inches
Courtesy of the artist

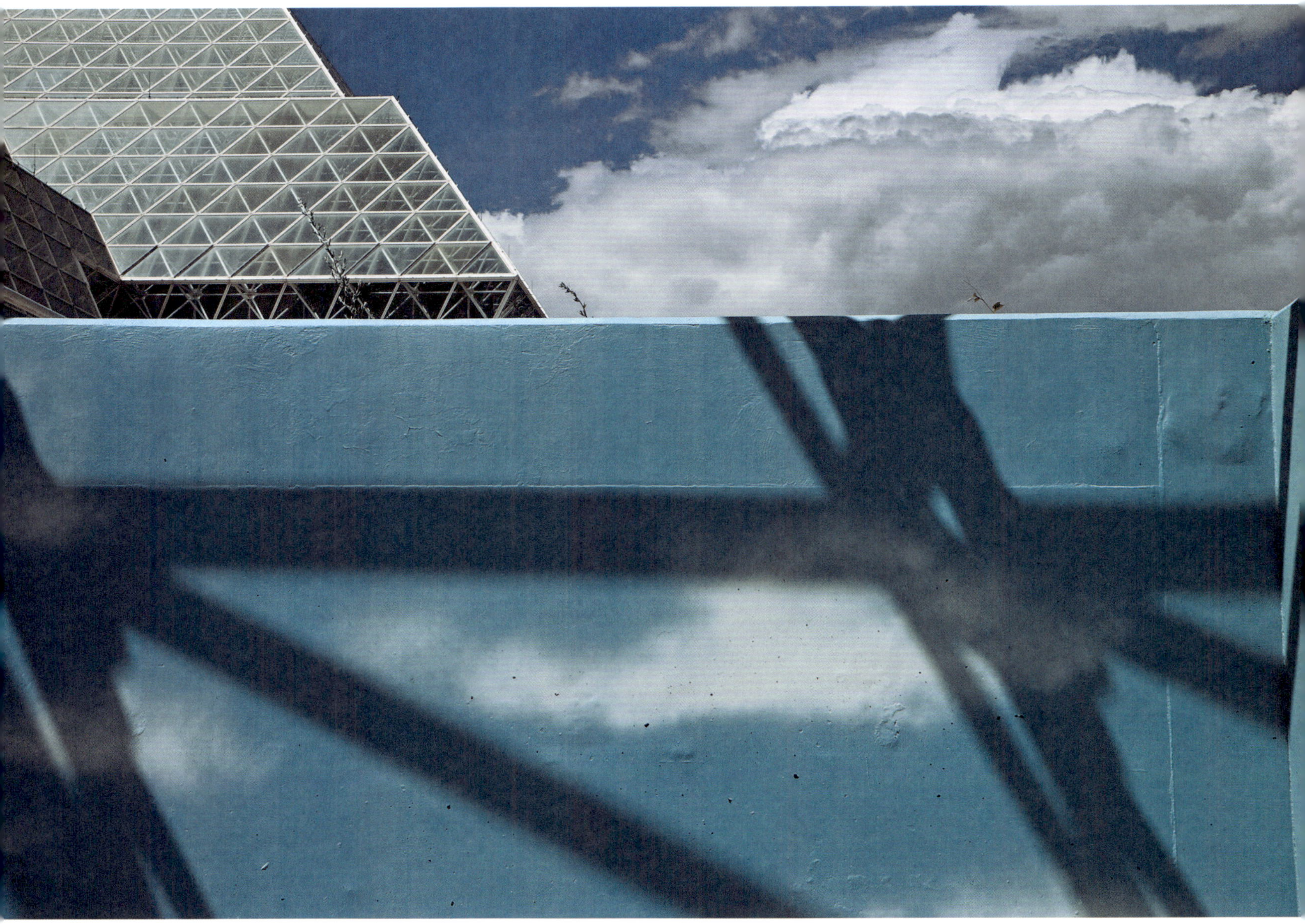

**DANA FRITZ**

***Painted Clouds, Biosphere 2*** **2007**
Archival inkjet print, 12 × 18 inches
Courtesy of the artist

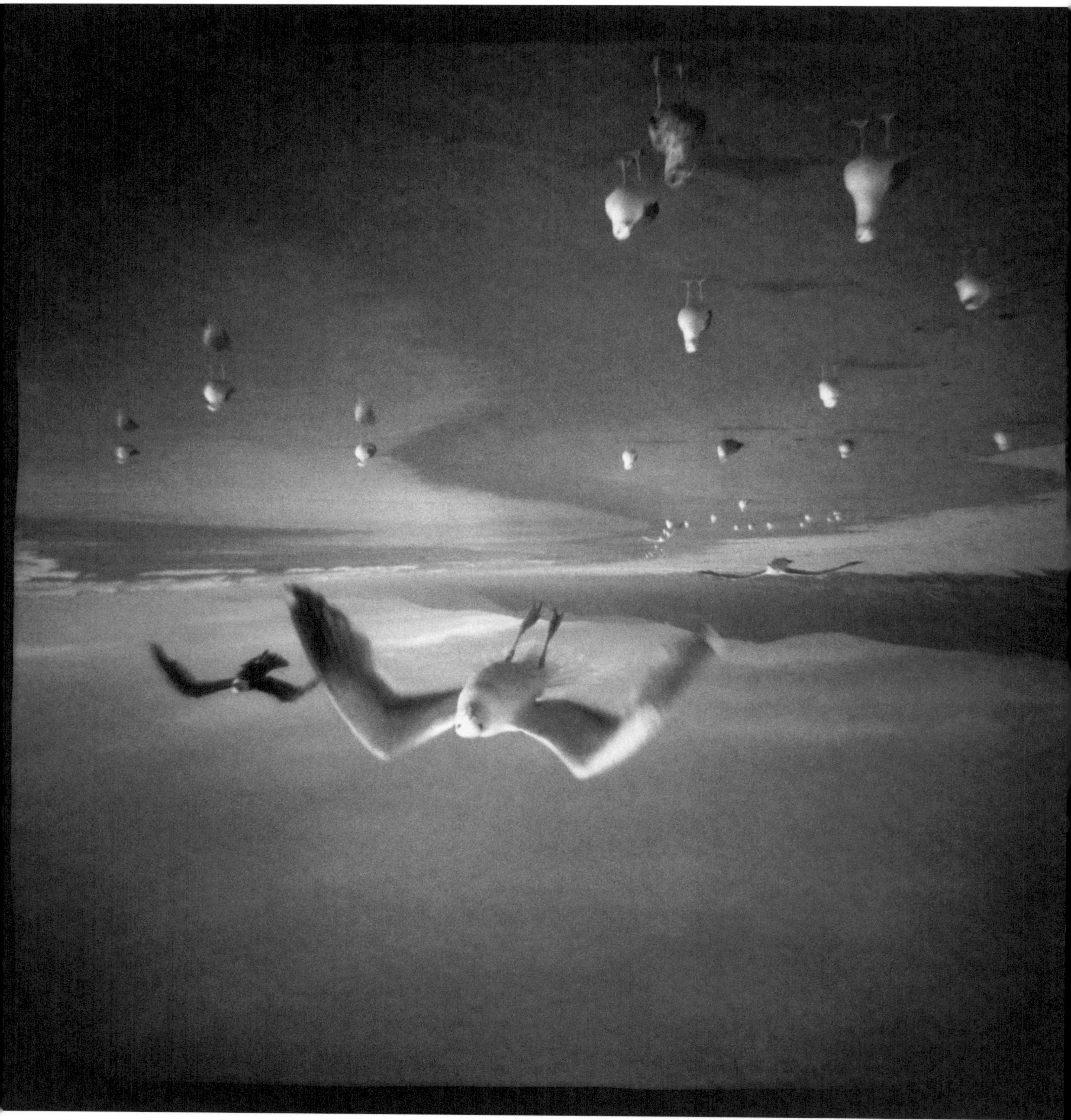

**KRISTA ELRICK**

***Oregon Gulls* 2005**
Silver photograph, pigment on paper, 13 × 13 inches
Courtesy of the artist

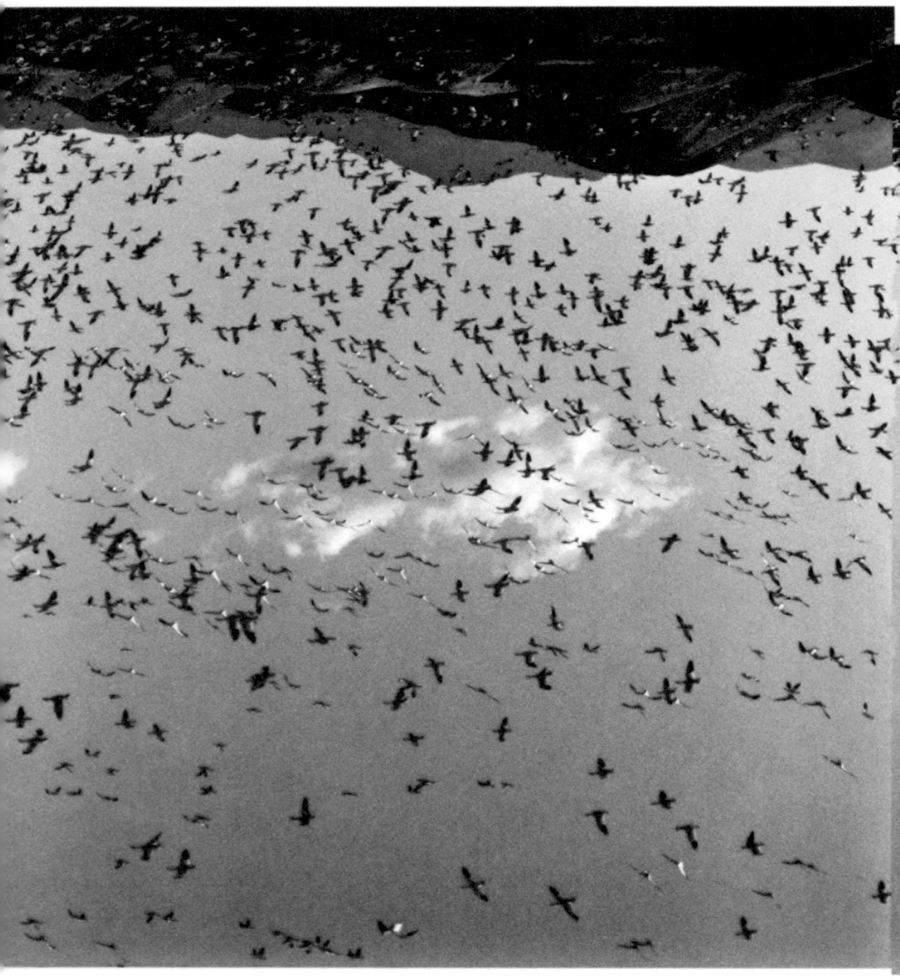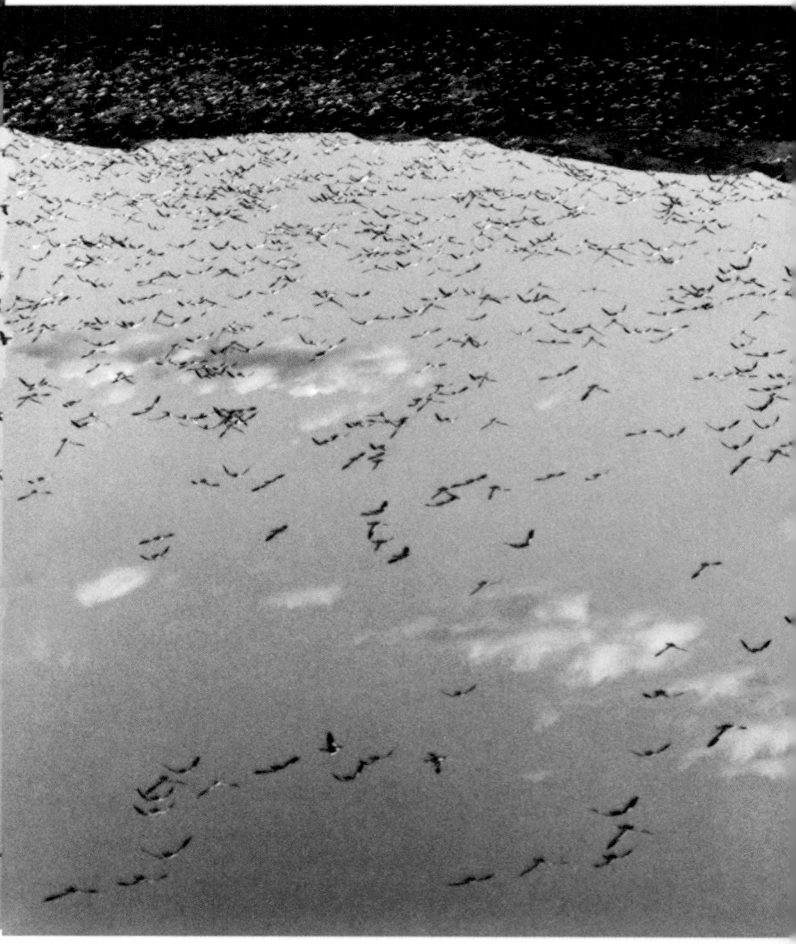

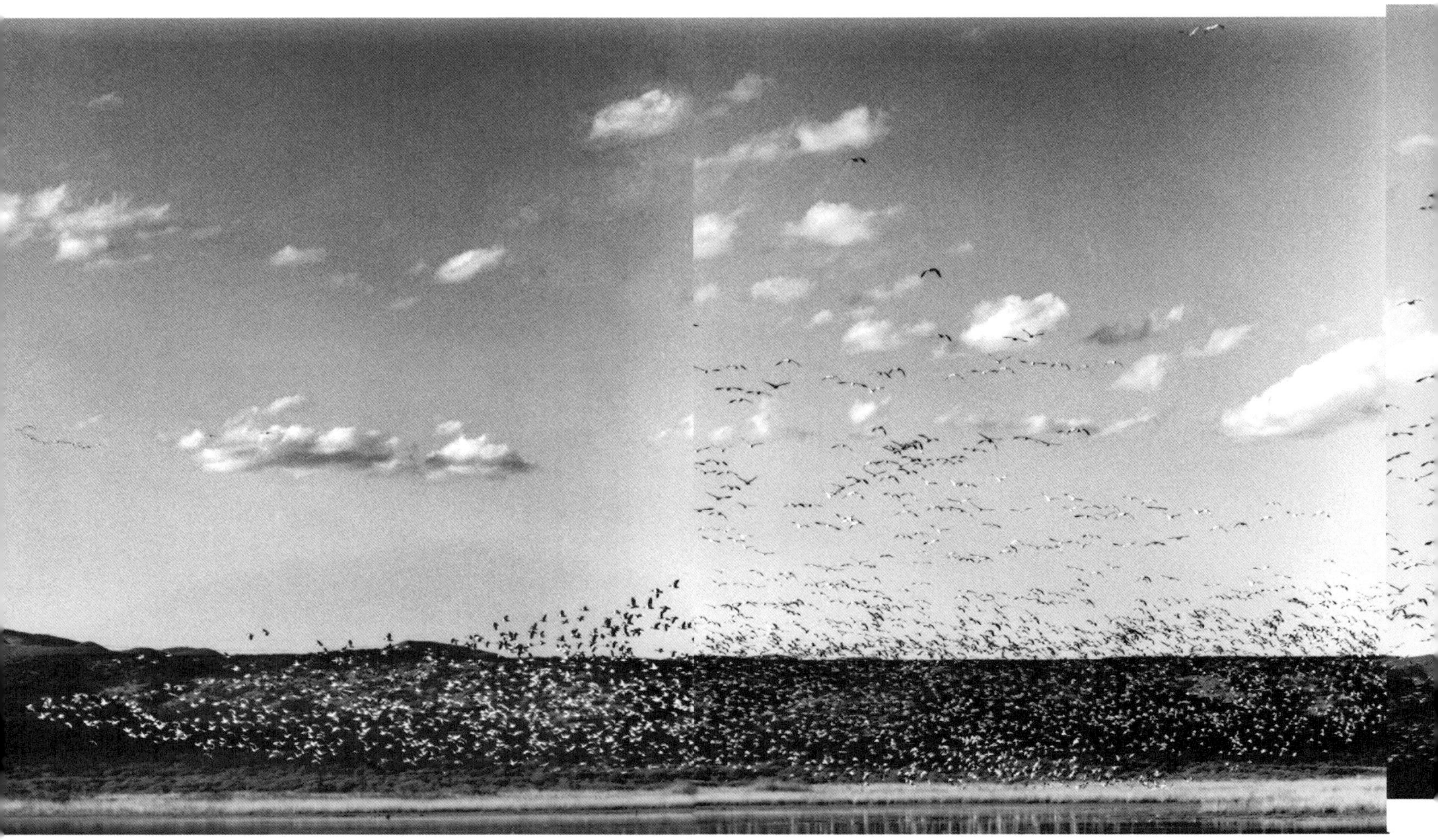

**KRISTA ELRICK**

***Snow Geese Flight* 2009**
Silver photograph, pigment on paper, archival laminate, 16 × 64 inches
Courtesy of the artist

# separating species

KRISTA ELRICK
DANA FRITZ
DAVID TAYLOR
JO WHALEY

stones in both US and Mexican history—points of disputed reference in subsequent international boundary disagreements and remnants of a storied past. As the new border fence is built, these monuments remain deeply present in contemporary psyche on both sides of the border.

**David Taylor** explores such issues with his series on the border monuments that demark the US/Mexico international boundary. Border monument number 1 marks the line drawn on the land between El Paso, Texas, in the United States and Ciudad Juarez, Chihuahua, Mexico. The monuments mark a sight line across the deserts of the Southwest until the final monument, number 258, ends between San Diego, California, in the US and Tijuana, Baja California, in Mexico. Six hundred ninety eight miles of mostly wilderness areas separate these two monuments. Taylor's goal is to photograph each monument from both sides of the border. Vantage point is critical to both his image making and to the interpretation of his photographs. Borderlines force an uneasy awareness of monuments and fences, of what it means to stand on one side of the line as opposed to the other side. More than halfway toward accomplishing his goal, Taylor is privileged: working for years with the border patrol, he has, until recently, been free to stroll at will back and forth across the border with his camera. Others are not so lucky. El Camino Real, Jornada del Muerto, El Camino del Diablo and the Devil's Highway are names given to the rugged historic migration routes, now blocked. The once permeable 2000-mile border will now have more than monuments to mark the international boundary: 15-foot pedestrian fences, vehicle barriers, radar, and surveillance cameras will monitor all comings and goings. The "Great Wall of Mexico," is nearing completion; ostensibly to halt illegal immigration, drug smuggling, and firearms trafficking crossing the border, the wall is designed to block coyotes, both human and animal, from crossing from one country to the next. My guess is that the coyotes will find other routes. In the meantime, many species will continue to die. At the Tijuana-San Diego border is another monument, dedicated to those who have died attempting to cross. Brightly painted coffins are mounted on the border fence. Each represents a year and the number of known dead. When habitats are destroyed, all species have three basic choices: adapt, migrate, or die.

Over the past 540 million years, the earth has experienced five great periods of mass species extinction. Nearly seventy percent of (often silenced) biologists and physicists view the present era as the sixth, and possibly the fastest period of mass extinction the planet has ever experienced. Some scientists predict that in the next one hundred years, humanity's pollution of the biosphere, overexploitation of species and natural resources, and human overpopulation could cause the extinction of half of all species currently in existence. Not surprisingly, back in 1950 when Native wisdom was ignored and the US Government eradicated the reservation's prairie dogs, the desert became a wasteland. Without the burrowing pups, the packed soil turned hard and impenetrable to the rains. Runoff over the parched earth created further erosion, eradicating the grasslands and leaving the sheep unable to graze. What the Navajo elders understood is that prairie dogs create community; when ecosystems, both in the desert and elsewhere, are destroyed, "there will be no one to cry for the rain." All things are intertwined: the rain, prairie dogs, grasslands, sheep, Navajo elders, birds, butterflies, environmentalists, scientists, writers, photographers, the Border Patrol, and even those in international government are part of a bigger ecosystem. Each has a part to play in understanding that the planet as a whole is better off without separating species.

pand, less free space remains for other species. Species that survive and thrive are those that will do well in a human-oriented landscape. Fritz asks us to consider the possibility that these places become "a substitute for less controllable and convenient outdoor experiences." She wonders "if replicas of threatened landscapes are an effective educational tool for motivating visitors to preserve the actual places they represent."

**Jo Whaley's** photographs bring the viewer to an artfully constructed consideration of the relationships between the human species and the world of insect species. Whaley considers the still-life sets she fabricates to be theatrical stages that create visual dialogue. She "embraces the concept that theater is the lie that tells the truth." Using butterflies and bugs along with "cast-off materials from urban production, which have been partially reclaimed by nature, including metals that have gone through fire, plastic that has been pitted by the sea, and paper that is foxed by microorganisms." Whaley's images point to the complicated interconnectedness that human have with the animal species of the planet. In her images, "natural specimens are combined with the detritus of our manufacture to mirror observations of the limbo between the natural habitat and the one we have altered, sometimes beyond the point of recognition. . . . The beauty of a butterfly's wing belies the reality that it is often toxic." In his book, *The Natural History of Butterflies*, scientist John Feltwell says, "no other animal is more typical of a healthy environment, or more susceptible to change than a butterfly." Twice each year monarch butterflies make their annual migrations from Canada and the US to the Oyamel forests in Central Mexico and back again. All three countries have established butterfly preserves to prevent the destruction of their natural habitat. Mexico takes a unique approach by providing the people who live near the preserves with jobs and income, thereby limiting logging and deforestation of the butterflies' wintering grounds.

The North American Commission for Environmental Cooperation, established under the North American Agreement for Environmental Cooperation to address environmental issues from a continental perspective, claims that grasslands are among the most threatened environments in North America, according to *Grassland Species of Common Conservation Concern*, the Commission's report from its first tri-national workshop. The main threat is habitat disturbance and loss. The continental grasslands were once the primary corridors connecting various populations of species throughout the Americas, including human as well as animal traffic. Wide-ranging grasslands link places together, maintaining the integrity of open spaces and protecting against fragmentation requires more than a local perspective. Diversity—the ability of natural and social systems to adapt to change—is essential for all aspects of life; it is the key to resilience. Biodiversity not only includes diversity in nature at the genetic, species, and ecosystem levels, it also extends to bio-cultural diversity–the diversity of societies, cultures, languages, and economics. Biological, socio-linguistic and socio-economic, as well as cultural diversity are the lifeblood of any sustainable human development and welfare. Problems affecting grasslands on the border between the US and Mexico have continental consequences to the habitat needs of all trans-boundary and migratory species.

Representatives of the United Sates and Mexico signed the Treaty of Guadalupe Hidalgo on February 2, 1848. The treaty ended the Mexican-American War and ceded nearly half of Mexico to the United States, creating a line across the continent that, in the words of one historian, "divided the indivisible." Article V of the Treaty established guidelines for determining the boundary. Surveying the capricious geography of the desert landscape in 1848 proved difficult at best. The obelisks and stone monuments of the boundary survey remain important touch-

## SEPARATING SPECIES: PHOTOGRAPHY IN THE ENVIRONMENT

MARY ANNE REDDING
Curator of Photography
Palace of the Governors/New Mexico History Museum

**"PRAIRIE DOGS CREATE COMMUNITY.** Destroy them and you destroy a varied world." Terry Tempest Williams recounts this story "In the Shadow of Extinction" written for the *New York Times* (February 2, 2007). "In 1950, government officials proposed getting rid of prairie dogs on parts of the Navajo Reservation to protect the roots of the sparse desert grasses, thereby maintain grazing for sheep. The Navajo elders objected, insisting that if you kill all the prairie dogs, there will be no one to cry for the rain." Globally, loss of habitat is the greatest threat to endangered species. Every year, an estimated 400 species of animals and plants go extinct in the wild. In the early 1990's Harvard biologist E. O. Wilson estimated that about 30,000 species go extinct annually. That's about three species an hour. The World Conservation Union now estimates that forty percent of all organisms are endangered. Two of the five species of North American prairie dogs, Utah and Mexican, are currently listed under the Endangered Species Act as threatened and/or endangered. Similarly, one-third of bird species in the United States are endangered, threatened, or in serious decline due to habitat destruction, pollution, and other environmental problems.

Recognizing that global warming, habitat loss, and fragmentation are the most serious threats to bird populations, the Audubon Society initiated the Important Bird Areas Program, a global effort to identify and conserve areas vital to birds and other biodiversity. **Krista Elrick** photographs important bird areas in various locations on the North American continent: Bonaventure Island in Canada, Alaska's Copper River Delta, and New Mexico's Bosque del Apache and the Galisteo Basin. Elrick is interested in the "migratory patterns of birds across contemporary natural, cultural, and political landscapes." She investigates both natural migratory settings and migration in urban areas. In a recent article about bird migration in the *New York Times Magazine* (February 22, 2009), author Jon Mooallem writes about "conservation-reliant species," which are dependent on humans for their survival. Not only is migration managed and protected by wildlife refuges, an intercontinental system of safe havens where birds can "stop, rest, and feed as they make their way from seasonal home to seasonal home," but biologists are also "donning hooded costumes and taking to the skies to lead birds on their annual migrations." Organizations like Operation Migration, Prairie Wings, and the Playa Lakes and Northern Great Plains Joint Ventures exist to establish, sustain, and enhance the international network of protected sites managed for migratory and resident birds and their habitats.

Sustaining the world's wildlife will require progressively more systemic interventions. **Dana Fritz's** ongoing photographic project, *Terraria Gigantica: The World Under Glass* explores three of the largest indoor landscape environments ever created: the controversial Biosphere 2 in the Santa Catalina Mountains north of Tucson, the Desert Dome and Lied Jungle at the Henry Doorly Zoo in Omaha, and the Eden Project in Great Britain. As Fritz says, all three human-made sites "function simultaneously as research laboratories and tourist destinations. Rich in historical precedents, these architectural and engineering marvels are working symbols of our current relationship with the natural world. Human attempts to cultivate non-native plants have a long history that parallels advances in engineering technology. . . . . [These] giant high-tech terrariums draw from this rich lineage of cultivating the exotic but also serve as laboratories for scientists to study everything from soil conditions to global climate change. The continued interest in facilities that contain and display plants and animals in simulated natural environments seems to indicate a palpable concern for nature as we know (or knew) it." As the human population and its influences continue to ex-

# separating species

KRISTA ELRICK
DANA FRITZ
DAVID TAYLOR
JO WHALEY